Sources and Studies
in the History of Mathematics and Physical Science

T0183855

Sources and Studies in the History of
Mathematics and Physical Sciences

Andersen K.
The Geometry of an Art: The History of the Mathematical Theory of Perspective from Alberti to Monge

Bos H.J.M.
Redefining Geometrical Exactness: Descartes' Transformation of the Early Modern Concept of Construction

Dale A.I.
Most Honourable Remembrance: The Life and Work of Thomas Bayes

Dale A.I.
A History of Inverse Probability: From Thomas Bayes to Karl Pearson

Dale A.I.
Pierre-Simon Laplace: Philosophical Essay on Probabilities

Damerow P., Freudenthal G., McLaughlin P., Renn J.
Exploring the Limits of Preclassical Mechanics

Federico P.J.
Descartes on Polyhedra: A Study of the *De Solidorum Elementis*

Friberg J.
A Remarkable Collection of Babylonian Mathematical Texts

Graßhof G.
The History of Ptolemy's Star Catalogue

Grootendorst A.W.
Jan De Witt's *Elementa Curvarum*

Hald A.
A History of Parametric Statistical Inference from Bernoulli to Fischer 1713-1935

Hawkins T.
Emergence of the Theory of Lie Groups: An Essay in the History of Mathematics 1869-1926

Hermann A., Meyenn K. von, Weisskopf V.F. (Eds)
Wolfgang Pauli: Scientific Correspondence I: 1919-1929

Hoyrup J.
Lengths, Widths, Surfaces: A Portait of Old Babylonion Algebra and Its Kin

Continued after Index

Ian Tweddle

MacLaurin's Physical Dissertations

 Springer

Ian Tweddle
Department of Mathematics
University of Strathclyde
Livingstone Tower
26 Richmond Street
Glasgow G1 1XH
UK

Sources Editor:
Jed Z. Buchwald
California Institute of Technology
Division of the Humanities and Social Sciences
MC 101-40
Pasadena CA 91125
USA

British Library Cataloguing in Publication Data
A catalogue record for this book is available from the British Library

ISBN-13: 978-1-84996-624-5   e-ISBN-13: 978-1-84628-776-3

9 8 7 6 5 4 3 2 1

Springer Science+Business Media, LLC
springer.com

Preface

Scotland had three important mathematicians who flourished during the first half of the eighteenth century and whose names are still revered in the mathematical world today. They are of course Robert Simson (1687–1768), James Stirling (1692–1770) and Colin MacLaurin (1698–1746). I have already been privileged to write about certain works of Simson and Stirling in earlier volumes in this Springer series (see [108, 109]). Now I am delighted to be able to complete a trilogy with this account of MacLaurin's MA dissertation and two essays for which he was awarded prizes by the Royal Academy of Sciences, Paris; these items are concerned principally with gravitation, collisions and the tides.

As on previous occasions I am indebted to many people and institutions for assistance, advice and encouragement. I would like to record my thanks to the following in particular:

my colleagues, Dr. Brian Duffy and Professor Ian Murdoch, who read early versions of parts of this work and provided guidance and information on some ideas from physics;

my colleague, Dr. Ronnie Wallace, our Departmental Computer Officer, for his assistance on many occasions with computing matters;

my former postgraduate student, Felipe Catalán, for discussions on the translation of the Corollaries which appear at the end of MacLaurin's MA dissertation;

an undergraduate student, Julie Lindsay, who elected to do her honours project (2005) on MacLaurin's essay on the tides, thereby concentrating my mind on producing for this purpose early versions of some of the material in Part III of this book;

an anonymous reviewer who made a number of valuable suggestions for improving presentation and developing various aspects of the work – I acted on most of these, ignoring only a few which would have taken me into areas where I did not feel competent to go;

the Department of Special Collections, Glasgow University Library, for providing me with much of my material, and for permission to use it for the present purpose;

my wife, Grace, and my son, Edward, for their general support and encouragement; and

my employer, the University of Strathclyde, for providing me with office, technical and library facilities.

MacLaurin's diagrams which appear in my Appendices II.2 (p. 80) and III.6 (pp. 209–210) were scanned from photographic plates provided by Glasgow University Library and are reproduced with permission. The source of the material in my Appendix I.1 (p. 31) is a hand-written note in the copy of MacLaurin's MA dissertation which belongs to the *Grace K. Babson Collection of the Works of Sir Isaac Newton* (on permanent deposit at the Dibner Institute and Burndy Library, Cambridge, Massachusetts); I am grateful for permission to reproduce this note. I worked originally with a very old, faint photocopy of this dissertation which is held in Glasgow University Library; Cambridge University Library kindly provided me with a modern replacement made from their copy of the dissertation.

Finally, it is a very great pleasure to record my appreciation of the work done on my behalf by the editorial and production staff of Springer-Verlag. I am particularly grateful to Mrs. Karen Borthwick, the Mathematics Editor at Springer-Verlag London, with whom I have worked on all three of my books in this series.

I.T.
Glasgow
September 2006

Contents

General Introduction

The main purpose of the present volume is to discuss, and to make generally accessible, three dissertations[1] by the Scottish mathematician Colin MacLaurin (1698–1746), who was regarded both in Britain and in continental Europe as one of the leading mathematicians of his time. These are:

(i)   his MA dissertation at Glasgow University (1713), which is largely concerned with gravity;

(ii) his essay on the collision of bodies, for which he was awarded the prize by the Royal Academy of Sciences, Paris, in their competition of 1724;

(iii) his essay on the tides, which earned him a share of the prize in the corresponding competition of 1740.

A brief review of MacLaurin's life and career[2] and of the plan of the present book will help to put these items in context and give an idea of the importance of MacLaurin's work generally.

MacLaurin was born in February 1698 at Kilmodan, Glendaruel, Argyllshire, where his father was a minister of the Church of Scotland. Much of the responsibility for MacLaurin's upbringing fell to his uncle Daniel McLaurin,[3] also a Church of Scotland minister, for his father died when Colin was just six weeks old and his mother died in 1707. In 1709 MacLaurin enrolled at the University of Glasgow with a view to studying for the ministry. However, he soon became fascinated by geometry, perhaps under the influence of Robert Simson (1687–1768), who was appointed Professor of Mathematics at Glasgow in 1711.

Following his graduation in 1713 MacLaurin appears to have returned to his uncle's home in Kilfinan, Argyllshire, where he continued to study mathe-

---

[1] I have used *dissertation* as a convenient word to apply to all three works. The term *essay* has usually been applied in the literature to items (ii) and (iii), although *treatise* is also found, as well as the French *mémoire* and the Latin *dissertatio*.

[2] The following account is based on my article [107]. For more detailed information see [54, 70 (73), 95, 104, 110].

[3] Various forms of the name are found: MacLaurin, Maclaurin, McLaurin, M'Laurin. The first form was certainly used by the mathematician (see facsimile of the last page of a letter of 7 December 1728 from MacLaurin to Stirling in [111], pp. 56–57) and I prefer to use it for him. However, see Appendices I.2 (p. 32), I.3 (p. 33), where McLaurin appears.

matics and divinity. However, in 1717 he became established as a mathematician by his appointment to the Chair of Mathematics at Marischal College, Aberdeen.[4] In 1719 he spent several months in London; there he met Newton and was elected a Fellow of the Royal Society. His first book *Geometria Organica* [65] was published in London in 1720, two related papers having been published previously in the *Philosophical Transactions* in 1718 and 1719.

During 1722–1724 MacLaurin served as tutor to the eldest son of Lord Polwarth,[5] one of the British ambassadors at the Congress of Cambrai. For a while they settled in Lorraine and it was during this period that MacLaurin prepared his essay on the collision of bodies. It seems that MacLaurin had left Aberdeen having neither sought permission nor made any arrangements for the conduct of his classes during his absence. On his return, following the death of his pupil, he was called to account but was apparently able to give an explanation acceptable to the authorities. However, their good faith was again abused: after some internal dispute, MacLaurin took up the post of assistant to the ageing James Gregory (1666–1742)[6] at the University of Edinburgh, leaving his superiors at Aberdeen to learn of his new position from the newspapers. Strong support from Newton had been an important factor in securing MacLaurin's appointment at Edinburgh. He remained there as Professor of Mathematics till the end of his life. In 1737 he was instrumental in forming the Edinburgh Philosophical Society,[7] which developed into the Royal Society of Edinburgh. His essay on the tides was communicated from Edinburgh.

The principal testimonial to MacLaurin's work is his *Treatise of Fluxions*, which was published at Edinburgh in 1742. It was designed in part to answer criticisms of Newtonian calculus: MacLaurin often provides two treatments of a topic, one using geometrical methods, the other based on the "modern" analytical approach. Here will be found, among a great many other top-

---

[4] Concerning MacLaurin's career at Aberdeen see [37, 90, 116, 117].

[5] Alexander Hume Campbell (1675–1740), Lord Polwarth 1709, second earl of Marchmont 1724 (see [54]). This son, Patrick Hume, died at Montpellier in 1724 from fever, much to MacLaurin's distress (see [70] ([73]), p. iv).

[6] This was the nephew of the great James Gregory (1638–1675). He retired in 1725 on a pension which was partly financed from MacLaurin's salary [41].

[7] Concerning this society and MacLaurin's part in its development see [36]. Amongst its publications are the *Essays and Observations, Physical and Literary*, also known as the *Edinburgh Physical Essays*, in which MacLaurin's papers [71] and [72] appear. The first volume (1754) contains in its Preface the following tribute to MacLaurin: "No sooner were public affairs composed [after the 1745 rebellion], than we met with an irreparable loss in the death of Mr. MACLAURIN, one of our secretaries. The great talents of that Gentleman are generally known and highly esteemed in the literary world; but the society have, also, particular reason to regrete in him the loss of those qualities, which form an excellent academician. Indefatigable himself, he was a perpetual spur to the industry of others; and was highly pleased with the promotion of knowledge, from whatever hands it came."

ics, detailed accounts of Taylor's theorem and its applications,[8] MacLaurin's contributions to the Euler–MacLaurin summation formulae, much geometry (including ideas of projective geometry), and versions or developments of the material contained in the second and third dissertations.[9]

MacLaurin was never one to avoid controversy: in the late 1720s and early 1730s he became embroiled in two quite public disputes, which perhaps damaged his reputation a little. The first of these was with George Campbell over priority in the discovery of certain results on complex roots of equations; to some extent Campbell was an innocent bystander, for MacLaurin's main grouse seems to have been that the Royal Society had published a paper by Campbell, having already published a paper by MacLaurin on the same topic, in which MacLaurin indicated that he would be providing a sequel (see [78, 115]). The second was occasioned by a little book on geometry and the description of curves published by William Braikenridge in 1733 [14]. MacLaurin claimed that he had shown some of the results contained in it to the author in the 1720s and accused him of passing them off as his own; one of the disputed results was a five-point construction of a conic, which MacLaurin certainly had in 1722 (see [67, 79, 106]). As we will see, there is a marked polemical thrust in the first two dissertations: Cartesian vortex theory is dismissed in favour of Newtonian ideas in the first and the Leibniz–Huygens concept of the "force of a moving body" being proportional to the square of its velocity is severely criticised in the second. Some of the results in the third dissertation overlap with the work of others, although MacLaurin himself appears not to have been involved directly in any dispute concerning priority on this occasion.

MacLaurin died at Edinburgh on 14 June 1746. In a sense he was a victim of the Second Jacobite Rebellion (1745–46).[10] Having been actively involved in arranging the defences of Edinburgh, he fled to York when it became clear that Edinburgh would fall to the rebels. He returned the following year in poor health, from which he never recovered. No doubt MacLaurin would have given much more to mathematics and natural philosophy had he lived longer. However, we do have the benefit of his posthumous *Account of Sir Isaac Newton's Philosophical Discoveries* [70] ([73]), which was "Published from the Author's Manuscript" and "Printed for the Author's Children." It has considerable relevance to the three dissertations under consideration here, for it provides background and descriptive material pertinent to all of their

---

[8] Contrary to what is asserted in many calculus books, MacLaurin was fully aware of Taylor's work, which he acknowledged in his book. As MacLaurin showed, the special case of expansion about 0 is sufficient for applications, for it just requires an initial change of origin.

[9] The nature and importance of the *Treatise of Fluxions* are discussed in [42].

[10] It is perhaps of interest to note that MacLaurin had benefitted from the First Jacobite Rebellion (1715). In its aftermath there was an almost total purge of the faculty at Marischal College; this included MacLaurin's predecessor, George Liddell, who was dismissed in 1716 (see [37, 116, 117]).

contents and it presents a mature and retrospective overview of Newtonian philosophy.[11] The work to be discussed below is founded very much on Newton's *Principia* and so it is relevant to note that for his first, second, and third dissertations MacLaurin would have had available to him its corresponding editions (1687 [81], 1713 [84], 1726 [85]).

In the following pages I have provided translations of the three dissertations. Each has its own introduction in which I have attempted to sketch the historical and scientific background to the work. Further comment, analysis and historical detail are given in the notes and appendices which follow each translation. The notes become increasingly more extensive and more mathematical as we move through the work; perhaps this reflects the importance of the individual dissertations: I have relatively little to say about the first, while my account of the third far exceeds the original in length. The presence and location of a note on a particular item is indicated in the individual contents pages and also by a page number in the margin beside the item; a note on a Proposition or Lemma also discusses any Corollaries or Scholia associated with the item. Occasionally a page contains footnotes from both MacLaurin and me; mine are numbered consecutively throughout the text, while those from MacLaurin are indicated by one of $*$, (a), (b), (c) (his own symbols). I have redrawn MacLaurin's figures in a form which I hope will make them more helpful: while following the originals as closely as possible, I have used dotted or dashed traces to separate out individual parts in the more complicated diagrams representing three-dimensional objects. MacLaurin's published diagrams are reproduced in Appendices II.2 (p. 80) and III.6 (pp. 209–210).

I first came upon MacLaurin's MA dissertation in the Manuscript Collection at Glasgow University Library. A photocopy of the original is contained among materials collected by J. C. Eaton (1915–1972), late of the University of Strathclyde, for a proposed but unrealised work on MacLaurin; also included is a draft translation by Eaton of this dissertation (MS Gen 1332).[12] The present translation (Part I) has been made by me independently of this, although I have compared parts of the two for general agreement. Since the original seems to be quite rare[13] I have reproduced the Latin text in Appendix I.4 (pp. 34–43). MacLaurin's sentence structure is quite complicated – perhaps he wanted to impress just as much with his Latin as with his scientific knowledge; however, I have tried to retain his style except where I felt that a literal translation was just too convoluted. The 15-year-old MacLaurin

---

[11] A detailed study of MacLaurin and Newtonianism will be found in [43].

[12] Concerning the importance of Eaton's scholarship and of his contribution to the development and expansion of higher education in Scotland, see the Foreword and Preface to [77].

[13] The *Biobibliography of British Mathematics and its Applications* [114] shows just two locations: Cambridge University Library and the Babson Institute. The latter copy, which is the source of the Glasgow photocopy, is now at the Burndy Library, MIT (see my Preface).

demonstrates in his dissertation[14] a confident understanding of Newtonian ideas on gravity and of the arguments against Cartesian vortex theory. He was later to make significant original contributions to many of the topics discussed in it.

Accounts of MacLaurin usually tell us that in 1724 (or 1725) he was awarded a prize for his essay on the 'percussion of bodies' but, as far as I am aware, no account has informed us about its contents. I have tried to fill this perceived gap with my translation and commentary presented in Part II. Apart from his nice geometrical treatment of oblique collisions, there is probably not a lot in this essay which is truly original; however, it does represent a powerful contribution to the bitterly contested dispute which was current at the time about whether the "force of a moving body" should be measured by its momentum or, effectively, by its kinetic energy.

The final dissertation, or essay (Part III), which is concerned with the tides, is certainly the most important. Its contents are rather better known, albeit indirectly, for a version of it is contained in the *Treatise of Fluxions*. However, the methods used there differ in some respects from those employed in the essay and the emphasis is rather on the related problem of the figure of the Earth, then a major topic of research, with the tides taking a subsidiary role. The most significant and original part of MacLaurin's essay is his discussion of the equilibrium of a fluid mass under the mutual attraction of its particles and certain external forces, which applies equally to the study of the tides and to the figure of the Earth. Some commentators have described MacLaurin's geometrical methods as "exact" (see, for example, [50]), in contrast to those of others such as Clairaut, who approximated the spheroidal Earth by a sphere. However, MacLaurin's treatment does involve many approximations, followed implicitly by limiting operations whose justification is a nontrivial exercise. Nevertheless, MacLaurin achieved a remarkable degree of completeness in his analysis, for which he fully merited his share of the prize. This aspect of MacLaurin's work was much admired by other mathematicians whose own developments in the area were often influenced by what he had done; such were Clairaut, Lagrange, Legendre and Laplace (see [103]) and in more recent times S. Chandrasekhar has given a rigorous account of "Maclaurin spheroids" in [24].

Colin MacLaurin certainly ranks among the greatest of the historically significant Scottish mathematicians, of whom there were not a few (Napier, Gregory, Simson, Stirling, MacLaurin, among others (see [41])). I hope that what I have attempted on the following pages will be seen as a worthy tribute to him.

---

[14]This dissertation, as well as the third, brings in some ideas from astronomy, for which the NASA website http://nssdc.gsfc.nasa.gov/planetary/ is a ready source of data.

Part I

MacLaurin on Gravity:

De Gravitate, aliisque viribus Naturalibus

(MA Dissertation, Glasgow 1713)

Part I Contents

Introduction to Part I

On 23 June 1713, at the age of 15, Colin MacLaurin graduated with the degree of Master awarded by the Faculty of Arts of the University of Glasgow [62]. As part of the requirements for this degree he presented and defended in public his dissertation, *De Gravitate, aliisque viribus Naturalibus*, which is the first item in this book. As far as I have been able to ascertain, candidates were not required to submit their work in printed form. It is likely therefore that MacLaurin had his dissertation printed with a view to its later use in the advancement of his career; indeed, it was probably submitted in support of his application for the Chair of Mathematics at Marischal College, Aberdeen, which he successfully contested in August–September 1717.[15] The dissertation is affectionately dedicated to MacLaurin's uncle, the Reverend Daniel McLaurin, who was a father-figure to his growing nephew (see Appendix I.2, p. 32).[16]

Much of MacLaurin's dissertation is devoted to promoting the Newtonian theory of gravitation and planetary motion and dismissing the rival vortex theory of René Descartes (1596–1650), which continued to attract powerful advocates (see below). MacLaurin also took up theological or philosophical aspects of his topic: Newton had identified a universal law by which the observed phenomena could be explained, but the question of why bodies attracted each other remained unresolved; for MacLaurin this was not a problem, but rather an effect to be ascribed to God.[17] Coincidentally, the second edition of Newton's *Principia* [84] was published in 1713; the date of publication is given as 11–14 July by Koyré and Cohen [63], so it seems most

---

[15]MacLaurin's dissertation was printed in 1713 at Edinburgh by Robert Freebairn, Printer to the King. The copy in the Babson Collection includes a hand-written note recording in Latin that MacLaurin had given the item to the writer as a gift when they were "the contestants for the vacant mathematical Professorship in the New College of Aberdeen" (see Appendix I.1, p. 31). The recipient must therefore have been Walter Bowman, the unsuccessful candidate [90].

[16]Daniel sent the dissertation for comment to his colleague, the Rev. Colin Campbell, minister at Ardchattan and a mathematician of some repute (see his entry in [54]). Campbell's enthusiastic reply, along with related correspondence, is contained among the Colin Campbell papers at Edinburgh University Library. A more accessible, typed copy of the reply is to be found among Eaton's papers at Glasgow University Library (MA Gen 1332, Box 1). See also Letter 116 in [77].

[17]See especially MacLaurin's Propositions VI, XII, XIV below.

unlikely that MacLaurin knew anything of the changes from the first edition when he produced his dissertation. It is well known that in the *Scholium Generale* added at the end of Book III Newton commented on the deficiences he found in the vortex theory and on the nature of God and his control of the system of planets and other celestial bodies.[18] There are many points of contact between MacLaurin's assertions and the views propounded there by Newton.[19] In common with the essays to be discussed in Parts II and III, MacLaurin's MA dissertation has a motto; in this case it appears on the title page and is appropriately a biblical quotation, namely, *Proverbs 3*, verse 19 (see Appendix I.3, p. 33): *The Lord by wisdom hath founded the earth; by understanding hath he established the heavens.*

An alternative source of inspiration for MacLaurin may have been the work of the cleric and scholar Samuel Clarke (1675–1729), who will also reappear in connection with MacLaurin's essay discussed in Part II. Clarke was a confidant of Newton's and had published in 1706 a translation into Latin of Newton's *Opticks* ([82, 83]). As Boyle lecturer, Clarke preached a series of sermons in 1704 and 1705 which were subsequently published in 1705 ([28], see also [30]); in *A Demonstration of the Being and Attributes of God* he presented and argued 12 propositions, or theses, of which the sixth is relevant here: "The Self-Existent Being, must of necessity be Infinite and Omnipresent." The same idea is developed in Newton's *Scholium Generale* and it has been suggested that either Clarke or Newton assimilated it from the other, or that possibly they formulated the concept through mutual discussion.[20] Clarke also produced in 1697 an annotated translation into Latin of Jacques Rohault's *Traité de Physique*, a classic of Cartesian physics; later he became critical of Cartesian principles and in the edition of 1710 presented Newtonian arguments against the vortex theories [93] (see also [94] and the Note on Proposition XIII below).

To be validated, a theory has to be capable of explaining the observed phenomena. In the case of terrestrial gravity, the immediate problem was to explain why bodies apparently fell towards the centre of the Earth. For planetary motions Kepler's laws provided the touchstone:

(i)   each planet moves in an elliptical orbit with the Sun at a focus;

---

[18]Roger Cotes, the editor of [84], also commented on these matters in his lengthy Preface. There seems to be some dispute about why Newton added the *Scholium Generale*. Was it just to defuse criticism of the *Principia* as a godless book? Had he come to believe that his science and theology were inextricably linked? Concerning such questions see, for example [56, 98].

[19]MacLaurin returned to the theological aspects in his *Account of Sir Isaac Newton's Philosophical Discoveries* [70] ([73]). See its Chapter IX, *Of the Supreme Author and Governor of the universe, the True and Living God.*

[20]See the article at http://en.wikipedia.org/wiki/Samuel_Clarke which was originally published in *Encyclopaedia Britannica* (1911). Newton had also touched upon theology in Queries 20 and 23 which were added at the end of Clarke's translation [83] (Queries 28 and 31 in [87]) and this material would no doubt have been familiar to MacLaurin.

(ii) the radius vector from the Sun to the planet sweeps out equal areas in equal times;

(iii) the square of the periodic time of a planet in its orbit is proportional to the cube of the major axis of the orbit.

The inverse square law of attraction, given by Newton but partly anticipated by others such as Robert Hooke (see [10], p. 98), was capable of explaining these and other phenomena and gradually it became accepted as the universal basis on which to investigate gravitational phenomena. It must be remembered, however, that vortex theory, first published in 1644, had attracted many distinguished physicists, who endeavoured with some success to make it work through the introduction of various modifications or to find a common ground on which both theories could be reconciled. Among such were Huygens, Leibniz, even Newton in his early work, and Johann and Daniel Bernoulli, who both received prizes from the Royal Academy of Sciences in 1734 for essays which used a mixture of Cartesian and Newtonian ideas. The rival theories were debated with particular vigour in France, some notable protagonists being Rohault, Malebranche, Mariotte, Roberval, Villemot, Saurin, Saulmon, De Mairan, Maupertuis, Molières and Bouguer. These are just a few of the distinguished names to be gleaned from the pages of E. J. Aiton's book [10], where the vortex and Newtonian theories are discussed along with their histories in considerable detail (see also Aiton's papers [6–9]).

In Cartesian theory, the planets float in a certain *subtle fluid*, also called *aether*, and are carried around in it by a system of rigid vortices centred on the Sun. Planets have their own individual systems of vortices in which their satellites circulate. To explain terrestrial gravitation it was asserted that the individual particles of aether, known as *boules*, circulate about the Earth more rapidly than the Earth rotates, as a result of which they have greater centrifugal force, therefore tend away from the surface of the Earth and in consequence a heavy body above the surface effectively sinks in the rising aether; arguments involving resultant forces were then needed to explain why the direction of downward motion is perpendicular to the Earth's surface rather than to its axis. This explanation is the topic of MacLaurin's Proposition II.[21] The Cartesian idea that gravitational effects are brought about by impulses from the particles of the circulating aether is taken up in Proposition V.

A further attack is made on vortex theory in Proposition XIII. Here MacLaurin points out that planets do not have circular orbits, as a simple vortex theory would require, and that attempts to allow elliptical rather than circular vortices lead to results which are not compatible with observations. Here he is following Newton (cf. [81] ([88]), Book II, Scholium to

---

[21]MacLaurin used the Latin term *thesis* in referring to the individual articles of his dissertation. I have translated this as *proposition*, its meaning in rhetoric and logic.

Proposition LIII; see also [10], p. 110): in the diagram below, aether circulating between the inner elliptical vortex and the outer, more nearly circular vortex would have to flow more rapidly as it crosses AD than it would when crossing CP, because of the narrower passage; consequently, in an elliptical vortex focused on the Sun S a planet would have greater speed at its aphelion A than at its perihelion P; this of course is inconsistent with observations and with the analysis based on the inverse square law of attraction. (See also my note on Proposition XIII, pp. 28–29.) The fact that the orbits of planets and comets do not lie in the equatorial plane of the Sun is also put forward as an argument against Cartesian theory.

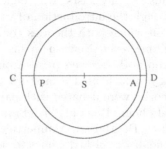

Kepler's second law is first mentioned in Proposition VIII and the third law in Proposition XI. The inverse square law combining both masses and distances is stated in Proposition XII, following discussion of the distance aspect in several of the preceding propositions. MacLaurin concludes in Proposition XIV that the heavens can contain no material capable of affecting the motion of the heavenly bodies in the way envisaged by the Cartesians. He observes in Propositions XV and XVI that all the heavenly bodies attract each other, so that their orbits are not determined by solar attraction alone; observed deviations from what is predicted by taking into account only solar attraction are to be explained in this way. The tides, which form the subject of the third dissertation[22] considered here, are touched upon in Proposition XVII.

The "other natural forces" mentioned in MacLaurin's title are examined in Propositions XVIII–XX. In these he is concerned with the cohesion of small particles in solids and fluids, the action of solvents, surface tension and capillary action, and the refraction, reflection and diffraction of light. There is a reference in Proposition IV to the attraction of magnets; essentially following Newton (Corollary 4 of Proposition VI in Book III of the *Principia* [81] ([88]), MacLaurin observes that gravitational attraction cannot be similar to the attraction of a magnet on a body. It is perhaps surprising that Newton is not mentioned until the final Proposition and then only in connection with optics. The dissertation ends with some rather obscure Corollaries, which, it may be presumed, express some philosophical and theological principles held by the author.

---

[22]Ideas of gravity are also reviewed in its Section II.

Translation of MacLaurin's Dissertation

A Philosophical Dissertation,

Concerning Gravity, and other Natural Forces.

I. Among the various phenomena of corporeal nature, there are two, which, as they are very greatly distinguished almost before all others, having been examined in themselves, have occupied to a very great extent the philosophers of all time. One of these is that *general tendency* towards its centre of all bodies moving about the surface of the earth, which is commonly called *gravity*; the other is the *regular gyration of planets* in their orbits, which recurs with definite periods. Various hypotheses have been devised by various people for the explanation in mechanical terms of those phenomena. An impartial examination of these will prepare the way for explaining and developing that general law of universal *gravitation*, to which, it will be established, those two most noble effects are to be referred as a common foundation, even if at first sight they seem to have nothing in common; from this we will also seize the opportunity to consider along the way certain other forces of nature, which it is necessary to put in place for the solution of certain other phenomena, which philosophers have undertaken to explain likewise by mechanical theories.

II. To make a start from the gravity of terrestrial bodies, the opinion of *Descartes* and of his followers deserves the first consideration. Among other wonderful effects which they invent for the *celestial matter*, they also derive gravity from its very rapid perpetual gyration about the earth; this gyration necessarily imparts to that matter a violent impulse away from the centre of the recessional motion, as a result of which terrestrial bodies, having much less force, are pushed down towards the centre of the earth. In this way water, or any other fluid, pushes upwards a body thrown into it which is specifically lighter. However, this hypothesis operates with the obvious disadvantage that it ascribes a really rapid, and even circular, motion to the *celestial material* (no traces of which present themselves to us in the nature of things), and to explain mechanically the origin and conservation of this material is a matter of equally great labour and effort, as to give an explanation of gravity itself.

Moreover, since necessarily this very material has to be supposed devoid of all gravity, what can nevertheless restrain its centrifugal impulse, which is continually so violent? It is not the pressure of another encompassing fluid, for it would be necessary for the former to be restrained in turn by the latter material, and for motion to be communicated to it; and since this fluid has to be supposed to be restrained by some other encompassing fluid, that will also restrain in turn: in this way it will come about that the motion of this material decreases continually when extended to infinity, and is finally reduced to nothing. Finally, since this material necessarily performs its orbits in circles parallel to the equator, it will be necessary for all heavy bodies to descend in the planes of those circles, and consequently in lines which do not tend towards the centre of the earth, but are perpendicular to its axis; this is entirely contrary to experience.

(p.27)   III. Others assert that *gravity* arises from the pressure of the overlying atmosphere, not noticing that the whole pressure of the atmosphere depends on this very gravity: for its elastic force itself, without some force acting against the elasticity, can cause no lasting pressure, since in this way the whole atmosphere would be rendered rapidly much thinner, as can be easily brought about in a pneumatic machine, in which we see clearly however that thinness of the atmosphere causes certain destruction to most, if not to all, animals. But this opinion is most effectively disproved from the fact that the force of gravity is found to be much more powerful, when the pressure of the atmosphere is removed, than when it remains: therefore it is so far from being the case that that pressure is the cause of gravity, that on the contrary it weakens the effect of this in all bodies, and in some it removes it completely: for it takes away from the gravity of any given body just as much as is equal to the gravity of the mass of air equal in volume to the given body; moreover, where a negative amount is left, as happens with bodies which are called *light*, the bodies do not descend but ascend.

IV. There are those who assert that *gravity* is an *attraction* of the same type as that by which a magnet attracts another magnet or iron; and consequently, if this can be explained mechanically, (which very many consider to be possible), the philosophical reasoning must apply equally to the former. However, a very brief comparison of both types of forces will show that the truth of the matter is quite different. As a result of the force of gravity the earth attracts in lines tending toward its centre, either exactly or approximately, any bodies which are moving round about it; and, as will be shown later, that occurs with forces which, at equal distances from the centre, are proportional to the amount of matter in the individual bodies, while at different distances they decrease in the ratio squared of the increased distances. A magnet, on the other hand, does not attract in this way towards its centre, but rather towards one or other of its poles; thus it does not attract equidistant bodies with forces proportional to their quantity of matter, so that in the case of equal bodies it attracts some with greater force, others with less force, and the greater part with none at all: in general it decreases in the ratio of the distances to a higher power than the square.

(p.27)   V. Not a few other arguments can be introduced, which overthrow sometimes only the proposed hypotheses, sometimes all other possible hypotheses, offering a mechanical solution of gravity; so far they certainly show that the descent of heavy bodies can result from no bodily impulse. In particular, since the momenta of the motions are always as the quantities of matter whenever the velocities are equal, and since heavy bodies at the same distance from the centre of the earth tend towards it with equal velocity (if we ignore the resistance from the atmosphere), it is clear that the impressed forces are directly as the quantities of matter in the bodies themselves, no account having been taken of the shape, texture or bulk. But if gravity were to arise from any impulse of a surrounding fluid, that impulse would consist either of a percussion of the parts, freely moved, of the fluid towards the same region to which the impelled body is driven, or of a pressure of the whole fluid pressing more powerfully against an obstruction placed on the other side: in the former case the force is impressed in proportion to the surface, in the latter in proportion to the bulk of the impelled body; in neither case is it in proportion to the quantity of matter. Moreover, every impulse pushes a body at rest to a greater extent than a body set in motion, so that, the greater the velocity by which the impelled body is moved, the less the impelling body adds increment of velocity to it, until the whole impulse stops, as well as the acceleration of the motion, the velocity of the impelled body and of the impelling body having been made equal: but gravity (as has been ascertained from very accurately set-up experiments) adds equal increments of velocity in equal time both to a very rapidly descending body and to a body starting at rest. It is therefore clear that gravity can arise from no corporeal impulse.

(p.27)   VI. If to the Proposition now proved two others are joined, it will be clear what is to be thought about the cause of gravity. One of these is as follows: suppose that a body placed at rest is moved from its position, the motion to be forced on it by some external cause, either corporeal or incorporeal; all the more so if the body, having been projected towards one plane, is cast back into the directly opposite plane; then that new and opposite motion is to be considered as resulting from an external cause. The other is that no body can move another body, unless by impulse, *i.e.*, a body can exert no force at a distance, in other words, it cannot act where it is not present. Therefore let us mention that, whenever we are following the commonly accepted and concise method of speaking of bodies *attracting* other bodies or *repelling* them without impulse, we wish to indicate by such phrases, not the true and properly named cause of the motion which is being discussed, but only the purpose for whose effect the force is applied for such movement in accordance with some general law of nature, and at the same time the boundary towards which, or from which, that force is directed: let it suffice to have advised of this once. The former shows that the gravity of terrestrial bodies arises from some external cause; the latter shows that its cause is not some corporeal thing, if indeed it is proved by the *above Proposition* that it does not arise from an impulse. Therefore it only remains for the cause of gravity to be recognised as a will capable of some incorporeal and intelligent cause which exercises its force uniformly according to a certain general law. But of what type this intelligent cause may be, will easily be accessible to anyone who considers that the whole structure of the globe of the earth is preserved and strengthened by this very gravity; otherwise this would rapidly fall to pieces, having been broken up by the centrifugal impulse. Gravity prevents mountains, seas, cities, people, and other living beings thrown off the surface of the earth from being scattered far through the vast region of the heavens. The subsistence and nutrition of both humans and the other living beings depend on gravity; thus it is that the lord of the earth and the preserver of mankind is to be recognised most deservedly as the creator of gravity.

(p.27)   VII. That the parts of the remaining planets and of the *Sun* are also joined together by gravity of this type is shown by their rotations about their axes, necessarily producing a centrifugal tendency, which would scatter rapidly those parts unless they were held together by gravity: indeed, these rotations in the *Sun* and very many of the planets are known through observations; moreover, in *Jupiter* especially they are known not only from the occasional gyration of the spots but also from the spheroidal shape arising from the same rotation, which is sufficiently discernable on account of the size of the body and the rapidity of the motion. Moreover, it will be clear from what is to be said later that this mutual gravity of the parts of individual planets towards each other agrees in all respects with our terrestrial gravity.

VIII. But the effectiveness of this principle is not contained within these boundaries; for a careful comparison of those effects will show quite clearly that that force by which planets are held in their orbits is certainly of the same type as that by which terrestrial bodies are pushed down towards the centre of the earth. It was demonstrated long ago that a body which is moved about another in such a way that, when radii have been drawn to the centre of the latter, it describes areas which are proportional to the times, is held in its orbit by a force which is constantly directed towards the centre of that other one. Therefore, since it has been determined that this is in fact the case with all primary planets and comets relative to the *Sun* and secondaries relative to their primaries, it is thus established that the force by which planets are kept in their curvilinear orbits has this in common with the gravity of terrestrial bodies: they tend towards the centre of some large body. Their agreement in other respects can be shown no less clearly.

(p.28)   IX. And first of all, it is proved as follows that the centripetal force of the *Moon* (by which it is pushed towards the centre of the *Earth*, as is clear from what has just been said) is the same as our terrestrial gravity. Gravity (according to very carefully set-up experiments with pendulums) drives terrestrial bodies down by $15\frac{1}{12}$ *Parisian* feet in one second, and thus (since the distances traversed by heavy bodies are as the squares of the times) by $60 \times 60 \times 15\frac{1}{12}$ feet in the first minute: in this same time the *Moon* is taken away from the tangent, to be diverted towards the *Earth* through a length of $15\frac{1}{12}$ feet: for a comparison of the periodic time and the size of the orbit shows clearly that the versed sine of the arc described in that time is of this size: therefore the accelerating force of the *Moon* towards the centre of the *Earth* is to the accelerating force of terrestrial bodies towards the same as $15\frac{1}{12}$ to $60 \times 60 \times 15\frac{1}{12}$, or as 1 to 60×60. And since the mean distance of the *Moon* from the centre of the *Earth* is sixty times the distance of terrestrial bodies turning about its surface from the same, it is clear that terrestrial bodies, as well as the *Moon*, are pushed towards the centre of the *Earth* by forces which are reciprocally proportional to the squares of the distances from the same. Further, since this is the nature of the centripetal forces of the *Moon* in the various parts of its orbit, being an ellipse described about the *Earth*, which is located at a focus, it is clear that terrestrial bodies as well as the *Moon* are pushed towards the centre of the *Earth* by the same force, varied according to the aforementioned law at the different distances.

(p.28)   X. Moreover, since this same law, namely, that centripetal forces are [reciprocally] as the squares of the distances, holds for all bodies describing some conic section about another point located at a focus, and since the orbits of all planets and comets are known to be of that type (if perhaps you exclude the *Jovian satellites*, whose perfectly circular orbits, if viewed separately, can be reconciled by means of some law of centripetal force), it is clear that the centripetal forces of them all are of the same type as is that force by which the *Moon* and terrestrial bodies are pushed towards the centre of the *Earth*.

XI. This same law of centripetal forces holds no less for different planets revolving about the same central body, as for the same planet at different distances from the body towards which it tends: in fact it has been demonstrated that, where several bodies revolve about the same central body in such a way that the squares of the periodic times are in the ratio of the cubes of the mean distances, they are all attracted to that central body by forces which are reciprocally proportional to the squares of the distances from the same body. Moreover, it has been ascertained from very accurate observations that all planets which revolve about the same central body, obey that very ratio of distances and times.

XII. Therefore, since the accelerating force of terrestrial bodies towards the *Earth* and of planets as well as comets towards their own central bodies decreases in the square of the ratio of the increased distances, this force will be equal in different bodies tending towards the same centre, at the same distance from it; and so their inertial forces, or weights, will be proportional to the quantities of matter in them. Moreover, since the reaction is always equal to the action, the tendency of that central body towards those other bodies will be equal to their weight, and so proportional to the amount of matter in them. It is therefore clear that universally the weights of bodies are in the ratio compounded of the direct ratios of the quantities of matter of the gravitating bodies and of the bodies into which they gravitate, and of the reciprocal ratio of the squares of the distances. And so, since the centripetal forces of planets and comets and the gravity of terrestrial bodies are clearly of the same type, there is no reason why we should not think that the former just as the latter are to be ascribed to the efficacious and uniformly acting will of the wisest and most powerful creator as the single cause.

(p.28)   XIII. Meanwhile the *Cartesians* undertook to solve mechanically this phenomenon, as almost all others: the refutation of their hypotheses must destroy all hope of a mechanical explanation. According to them, by rotating about its axis, the *Sun* carries around a certain subtle fluid and the primary planets, which are swimming in it; these also have individually their own vortices, in several of which the secondaries are carried away. But first, since planets do not describe circles, they cannot be carried around in vortices which are infinitely extended or confined by a spherical vessel; but if the bounds of a vortex are arranged otherwise, the planets will deviate more from a circular path the further they are from the centre; and the aphelia of them all will be found in the same celestial region: for otherwise the eccentricity of the lower planets would be much greater than that of the higher planets; the aphelia of *Mars* and *Venus* would be almost opposite; for their distance at the beginning of *Virgo* is almost one and a half times the distance of the same at the beginning of *Pisces*. This observation provides another argument against the vortex hypothesis. For, since the motion of a fluid carried around through unequally sized canals must be more rapid in narrower places, it is clear, according to the *Cartesian* hypothesis, that a fluid in which the *Earth* is swimming (and therefore the earth itself), intermediate to those two orbits, must be carried more rapidly at the beginning of *Pisces* than at the beginning of *Virgo*: this is clearly incompatible with observations. Furthermore, if the vortices are homogeneous, the periodic times will be as the squares of the distances; but if they are heterogeneous and the parts further away from the centre are more dense, as *Descartes* maintained, and the theory requires, the periodic times will be as some higher powers of the distances; however, the periodic times of the planets are only in the ratio of the mean distances, raised to the power one and a half. But the *Cartesian* vortices are most effectively rebutted by the inclination of the planetary orbits to the axis of the *Sun* and to one another, and by the motion of the comets, at one time directly opposed to the movement of the planets, at another time perpendicular to their orbits.

(p.29)   XIV. Therefore, since the *celestial material* (if there is any) is not carried around with the planets, and besides it will not have impeded their motion to any noticeable extent over so many thousands of years, and since it opens up such an easy way for comets swimming very rapidly through it, it is clear that the regions of the heavens are as free as possible and consequently no material, which is sufficient to deflect regularly the continuous motion of so many bodies, is to be found in them. Therefore the motion of planets and comets in curvilinear orbits arises from no impulse of any imperceptible small bodies, and so from no mechanical cause. And hence it adds much to the magnificent idea, established by the *6th Proposition*, of the creator of gravity, who, it is now agreed, is master not only of the whole earth, but also of heaven, and the protector of all its inhabitants; who preserves the structure of all heavenly bodies; by whose powerful right hand the planets, driven in perpetual orbits about a common central body, are saved from being perpetually frozen and enveloped by the densest darkness and losing all other things which are concerned with the preservation of vegetation or animals as a result of having been carried away by a centrifugal impulse through the empty vastness and deprived of every benefit which they now receive from the *Sun*. But, just as this centrifugal impulse, if it were not restrained by gravity, would cause to all planets certain damage, by carrying them away from the *Sun*, so no less would gravity bring upon them certain destruction, by casting them into the burning atmosphere of the *Sun*, if projectile motion had not been impressed upon them: indeed, when these two forces have been combined, it is necessary that they are carried about the *Sun* in some curved line; this line will be circular if the direction of the projectile motion is perpendicular to the radius drawn to the *Sun*, and its force will be equal to the force of gravity: but if either of these conditions is lacking, that curve will be some conic section. The things that are said here concerning the primaries with respect to the *Sun*, are to be understood for the secondaries likewise with respect to their primaries.

(p.29)   XV. Now it has been shown that the primary planets gravitate towards the *Sun* and the secondaries towards their primaries: moreover, since any body which describes about another, however it is moved, areas which are proportional to the times, is driven by all the accelerating force by which that other is driven in addition to the force tending towards that other, it is thus clear that the secondary planets, no less than the primaries, are heavy towards the *Sun*. But it is clear from certain perturbations of their motions, which can be derived from no other cause, that mutual gravitation affects not only the primary planets in connection with the *Sun*, and the secondaries in connection with the *Sun* and their primaries, but also planets of the same order, *e.g.*, the primaries among themselves; such effects are the migrations of the *apsides* and the *nodes*, etc., which are quite perceptible whenever they reveal themselves, especially those in *Jupiter* and *Saturn* round about the heliocentric conjunctions of these planets, on account of their vast size and distance from the *Sun*, and the simultaneous slowness of their motion. Moreover, since the motions of their *satellites* are also found to be perturbed perceptibly in those conjunctions, it is clear that there is also an interaction of gravitation between the primary planets and the secondaries of others.

(p.29)   XVI. Indeed nothing demonstrates more clearly the effectiveness of that universal law, according to which all bodies gravitate mutually towards each other, than those variations which have so racked the minds of astronomers of all time, namely, the irregularities of the lunar motion. For, if the law of gravitation is assumed, the accelerating force towards the *Sun* of the *Moon*, whose distance from the *Earth* is of significant magnitude (even when it is compared with the distance of the *Earth* from the *Sun*), must sometimes be greater and sometimes less than the accelerating force of the *Earth* towards the *Sun*: this inequality will be greatest when the *Moon* is in the syzygies; in the quadratures it will be least, or there will be none at all; as a result of this it turns out that its motion from the quadratures to the syzygies (other things being equal) is accelerated, while that from the syzygies to the quadratures is retarded; and so the curvature of its orbit and the distance from the primary (other things being equal) will be greater in the latter, than in the former: hence the *Moon* also does not always describe areas about the *Earth* which are exactly proportional to the times: these things all agree very well with observations. In a word, whatever irregularities in the motion of the *Moon* are detected by observations (indeed very many are detected), they are explained *a priori* as a necessary consequence of the assumption of what we have called the universal law of gravitation, which is therefore to be considered as corroborated to a very great extent by them. Also from the same law, and with equal clarity, is deduced the known *precession of the equinoxes* and the *oscillation of the axis of the Earth*, which takes place twice a year.

(p.30)   XVII. Moreover, according to this law, the parts of any terrestrial fluid gravitate towards the *Moon* or the *Sun*, perceptibly more when turned directly towards the *Moon* or the *Sun*, but less when turned away, than the centre of the *Earth*, or its whole mass taken together; in this way such an amount is consequently taken away from their gravity towards the *Earth*: however something is added to the gravity towards the *Earth* of the parts which are lateral or intermediate between the averted and obverse parts, when the attraction of the *Sun* or of the *Moon* acts together with it a little: hence it follows necessarily that, while the averted and obverse parts are lighter, the lateral parts on the other hand are heavier, the former having been pressed upwards by the latter, until they counterbalance through the height of the columns, because there is a deficit in their accelerating gravity: moreover, the forces of the *Sun* and of the *Moon* bring about the rise of the terrestrial fluids (namely, the atmosphere and the sea), which is not a twofold effect but a unique one to be determined from their combination; because of the different distances of those luminaries from the *Earth* and their declinations from the equator, this must vary, namely, in the cube of the inverse ratio of those distances. And from this fact, and no other, all phenomena of the tide of the sea can be very easily deduced; these things therefore bring the greatest confirmation to the principle of gravitation, which has now been validated.

XVIII. In addition to that gravity, which we have been discussing so far, by which all particles of matter tend mutually towards each other, without any distinction of shapes, forms, circumstances or motions, the forces decreasing in the square of the ratio of the distances, there is also a certain other force, by which very small particles of matter which touch each other, or are very close to contact, tend mutually towards each other more powerfully than according to the law of gravitation just explained: this force is reduced in more than the square of the ratio of the increased distance: and since this force acts only where there is contact or almost contact, the cohesion of any two particles of matter will be stronger according as their contact is greater; and so particles which have larger surfaces which are flat, or at least mutually congruent, adhere very firmly to each other; but those which are spherical, or else have convex surfaces, adhere more weakly (if at all); particles of the former type make up a moderately hard body, while those of the latter type form a fluid; and from the various intermediate types of contact arise various cohesions: thus, in a word, otherwise unsolvable phenomena, both of *solidity* and *fluidity*, can easily be explained. But since this force reveals itself not a little at very small distances, although the parts of the body may be separated somewhat by some external force, as long as they coalesce no more closely with new particles, then, when that external force has been taken away, they will revert to their former contacts and cohesions; in this way the body will recover its former shape, which otherwise it will necessarily lose completely. And the nature of *elasticity* and *flexibility* is very well explained in this way. And from these things it can be understood how a great difference of attractions arises in different particles as a result of their different shape and texture; for on this account some things tend mutually towards each other with scarcely any force, while others do so with very great force: most notable among the latter are the acid salts which generally predominate in solvents; for, having been attracted by the particles of the body to be dissolved, they fall down into its pores, as long as they are of suitable size, with such a large force that they separate the particles unless they stick together very strongly. The solutions of all bodies are easily explained in this way.

XIX. Also from this mutual attraction of very small particles of matter, an exceedingly large number of phenomena of fluids, which would otherwise be unsolvable, can be easily explained. For, from the fact that particles of water attract particles of wood or glass more powerfully than each other, arises that known phenomenon that water confined in a wooden or glass vessel is higher near the sides of the vessel than in other places; and so in very small tubes immersed a little in it the water is higher inside the tube than outside; but since particles of mercury attract each other more powerfully than particles of wood or glass, the effect is quite the reverse. Hence it may also be that, since they must fall under the force of gravity, tiny drops of water and of other fluids are propped up by glass, wood and very many other bodies. And just as a spherical figure of the planets necessarily arises from the equal gravity of the parts in the planets mutually towards each other, so from the equal centripetal force of the particles of water, mercury and similar fluids, mutually approaching right up to each other, arises the spherical shape of tiny droplets in those fluids. From what has just been said the reason for the congruity of water with wood, glass and other bodies and the incongruity of mercury with the same bodies can be understood; and with equal ease all remaining phenomena of the congruity and incongruity of fluids are resolved. Finally, it is clear from this why grains of sand and several other tiny bodies which are specifically heavier than water nevertheless do not sink in it: namely, the mutual attraction of particles of water, although it may be very small, nevertheless produces some resistance, to the overcoming of which the gravity of those small bodies is not equal. Most of these phenomena were explained by very many people through the action of the atmosphere; their error is shown by the fact that these phenomena are also found to occur in a vacuum.

(p.30)   XX. The same mutual attraction of the particles of matter having been supposed, the phenomena of crystallisation, precipitation, the congealing of fluids, electricity, and very many others can be explained very easily; it is not possible to dwell on these matters. But the explanation of the refraction of rays of light, which comes out from this, is a matter more worthy than something which deserves to be passed by completely untouched. Tiny particles or rays of light are bent in their passage near the corners of bodies (as is confirmed by the observations of the distinguished *Newton*), the effect being greater the nearer they approach to the bodies: it is quite clear that such a regular bending results from no impulse of particles flowing out from the bodies, but from some completely unmechanical force, which is impressed upon them by the Author of nature, according to a certain law, in proportion to the various distances from the bodies, towards which they are directed, or from which they are receding: a force of this type having been assumed, the author who has just been extolled has shown that it necessarily follows that the sine of refraction is always in a given ratio to the sine of incidence, whatever the obliquity of the incidence; this he shows to be the case by experiment. It is therefore necessary that the rays which fall obliquely from a rarer medium into another, which is more dense or in some way more attracting, having been attracted by this denser medium, are bent before they come in contact with it, so that the line of direction of the ray after it has entered the body makes a smaller angle with the perpendicular than before the bending: and hence comes about refraction towards the perpendicular. But if the ray of light falls obliquely from a denser medium into a rarer one, or one that is at least less attracting, then, on account of the greater attraction of the former, it will be curved towards it on or immediately after exit, so that the direction of the ray now makes a greater angle with the perpendicular than before: and hence comes about refraction away from the perpendicular. But, if in this case the angle of incidence is exceedingly large, refraction will be changed into reflection, so that the angle of incidence is equal to the angle of reflection: it is clear that meanwhile the motion of a particle of light is accelerated in the former case but retarded in the latter; and hence it is that the velocity of light is generally much greater in a denser medium than in a rarer one. Moreover, when the ray is directed only towards the parts lying perpendicularly below it, clearly the ray stays in the same plane perpendicular to the refracting surface throughout the whole period of bending. Again, from the different forms of the rays of light, or perhaps from different velocities, there arise different attractions among the rays of light and some bodies and so different degrees of refractability. Also by some similar principle may be explained those amazing alternations of easier reflection and transmission, which the same most distinguished author has shown by very many experiments to occur in rays of light.

COROLLARIES.

I. The simple and unordered nature of the mind does not allow it to exist in any part of space in such a way that it is coextensive with it; nor indeed does it prevent it from being present in one place, namely, where the body is, in such a way that it is not present similarly in another place.

II. Although the real or absolute essences of substances are unknown to us, it in no way follows from this that we can pronounce nothing certain concerning their dispositions and mutual relationships.

III. Moral philosophy rests as it were on the necessary foundation of the existence and providence of the greatest divine power, especially in so far as this reveals itself in the dispensing of rewards and penalties.

IV. For the sake of preserving life or averting some serious injury, any laws can be set aside, namely by actions indicating agreement, even if extorted by the very unjust ferocity of the one in whose favour they are put forth, until they are set aside; thus it can happen that as a result of such action one man does not have the right to seek anything or to keep something in his possession, while another, who has willingly committed himself to bringing some law to the matter and thus to taking an obligation upon himself, is bound entirely by trust.

THE END.

Notes on Part I

Note on Proposition III (p. 16). Newton makes similar observations in the *Scholium Generale*, referring more specifically to "Mr. *Boyle's* vacuum." At the end of the article MacLaurin applies the principle of Archimedes: the body experiences an upthrust equal to the mass of the displaced air.

Note on Proposition V (p. 17). Concerning collisions, which are the topic of the second dissertation discussed in this book, see my Introduction to Part II, p. 49. For a body falling under gravity we have $v = v_0 + gt$, where v_0 is the velocity at time $t = 0$; this is independent of the mass of the body and leads to $v_2 - v_1 = g(t_2 - t_1)$; thus, as MacLaurin asserts, there are equal increments of velocity in equal times irrespective of the initial velocity.

Note on Proposition VI (pp. 17–18). The statement that "a body can exert no force at a distance, in other words, it cannot act where it is not present" was a much debated philosophical tenet. MacLaurin uses it to assert that gravitational attraction cannot be an impulsive force. But it could also be seen as denying the possibility of one body attracting another body at a distance from it. It is perhaps of interest to note that the theological aspects of MacLaurin's Proposition VI share common ground with the ideas expressed in Clarke's sixth proposition, to which reference was made above.

Note on Proposition VII (p. 18). It is now believed that Jupiter's spots are caused by vortices, in the sense of cyclones and anticyclones in its atmosphere, which are visible because of the material stirred up by them. These storms can last for centuries, although it appears that in recent times spots have been combining or even disappearing. Jupiter has a marked oblateness due to its liquid composition and rapid rotation about its axis. About 1740 Maclaurin presented a paper to the Edinburgh Philosophical Society concerning the surface of Jupiter; this was published posthumously in 1754 [72]. In it he speculates on the possibility of very large tidal effects on Jupiter brought about by the conjunctions of its four known moons. He also observes that, since Jupiter is noticeably spheroidal in shape, it is definitely wrong to consider that particles gravitate towards its centre.

Note on Proposition IX (pp. 18–19). Here MacLaurin is following Newton's discussion in Proposition IV of Book III of the *Principia* [81] ([88]). A Parisian foot was about .325 metres, which makes $15\frac{1}{12}$ Parisian feet about 4.9 metres. The distance travelled by a body falling under gravity is given by $s = \frac{1}{2}gt^2 + v_0t$, where v_0 is the velocity at time $t = 0$. Thus, a body falling under gravity from rest travels $\frac{1}{2}g$ units in the first second, where g is the acceleration due to gravity; with $g = 981\,\text{cm/s}^2$ we have approximately the figure quoted by MacLaurin.

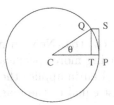

The *versed sine* of the arc PQ of the circle shown is the length TP; in modern notation this is $R(1 - \cos\theta)$, where R is the radius of the circle. It is also equal to the length QS, which is what MacLaurin means by the diversion from the tangent. The mean distance of the Moon from the Earth is 384,467 km and the Earth's volumetric mean radius is 6,371 km, which is consistent with the assertion by MacLaurin (from Newton) of a factor of 60.

Note on Proposition X (p. 19). The word *reciproce* [reciprocally] was written in by hand in the original. The Jovian satellites to which MacLaurin refers are the Galilean satellites *Io, Europa, Ganymede* and *Callisto*, which were discovered by Galileo in 1610. Their orbital eccentricities are 0.004, 0.0101, 0.0015 and 0.007 respectively, and so it would have been difficult in MacLaurin's time to distinguish their orbits from circles. By way of contrast let us note that for the Earth and the Moon we have 0.0167 and 0.0549 respectively for the orbital eccentricity.

Note on Proposition XIII (p. 20). MacLaurin refers here to the zodiacal calendar in which dates correspond to when the ecliptic crosses certain constellations: in the present-day *astronomical* version the beginning of Virgo is about 17 September and the beginning of Pisces is about 12 March. He uses an argument which is an extension of that cited against the vortex theory in my Introduction (pp. 13–14): the orbit of the Earth is intermediate to the orbit of Venus (nearer the Sun) and the orbit of Mars (further away from the Sun); if the Earth is carried around by aether circulating between the vortices of Venus and Mars, then the speed of the Earth will be greatest where these vortices are closest and least where they are furthest apart; again this is not consistent with observations. This material appears to be drawn from Newton's Scholium to Proposition LIII in Book II of [81] ([88]); it is also found in Clarke's notes in [93, 94] (see his note on Part II, Chapter 25, Article 22). MacLaurin's point about eccentricities may be based on the following: in a model in which each orbit has a focus at the Sun, semimajor axis of length $a + d$ and semiminor axis of length $b + d$, it is easily seen by elementary calculus that the eccentricity decreases as d increases.

Note on Proposition XIV (pp. 20–21). The term *projectile motion* refers to the initial conditions under which a planet was set in motion, its orbit being determined thereafter by the inverse square law of attraction.

Note on Proposition XV (pp. 21–22). MacLaurin refers to the "helio-centric conjunctions" of Jupiter and Saturn: these occur where the planets are aligned with the Sun and are on the same side of it. Then the attraction of the Sun on Saturn, the outer planet, is augmented by the attraction of Jupiter, the inner planet, on Saturn, while the attraction of the Sun on Jupiter is diminished by the attraction of Saturn on Jupiter. Significant forces are involved here: the distances from the Sun to Jupiter and from Jupiter to Saturn are comparable and these planets are massive, the mass ratios of Jupiter and Saturn to the Earth being 317.83 and 95.159, respectively. The *apsides* of an orbit are the points where the orbiting body is closest to or furthest from its central body; the *nodes* are where the orbit crosses the ecliptic. These vary due to the influence of other orbiting bodies.

Note on Proposition XVI (p. 22). Here MacLaurin comments on the distortion of the Moon's orbit about the Earth due to the attraction of the Sun. In the syzygies the Sun, Moon and Earth are aligned, with the Sun and Moon in conjunction (Sun–Moon–Earth) or in opposition (Sun–Earth–Moon); in the quadratures the radius vector from the Earth to the Moon is perpendicular to that from the Sun to the Earth.

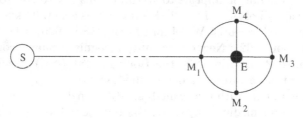

In the quadratures (M_2, M_4) the distances SM_2, SE, SM_4 are approximately equal, so the accelerations of the Earth and the Moon towards the Sun are approximately the same. However, in the syzygies (M_1, M_3) the acceleration towards the Sun of the Moon at M_1 will be greater than that of the Earth and at M_3 less than that of the Earth. Relative to the Earth the motion of the Moon will be affected by an accelerating force towards the Sun as it moves from quadrature to syzygy M_1 and then there will be a retarding force as it continues on to the next quadrature.

Newton discussed the precession of the equinoxes in Propositions XXI and XXXIX of Book III of the *Principia* [81] ([88]). By "oscillation of the axis of the Earth" MacLaurin probably means the nutation of the axis referred to by Newton in Proposition XXI: "the axis of the earth, by a nutation in every annual revolution, twice vibrates towards the ecliptic, and as often returns to its former position" [15].

Note on Proposition XVII (pp. 22–23). The roles of the Moon and the Sun in bringing about the tides in the waters of the Earth, which is considered in this Proposition, is of course the subject of the third of MacLaurin's dissertations and more detailed discussion will be found in Part III. The clause "until they counterbalance through the height of the columns" refers to a known condition for the equilibrium of fluids: columns through the fluid meeting at a point must exert equal pressures at the point. Concerning MacLaurin's reference to "the cube of the inverse ratio of those distances," see my Introduction to Part III, p. 91.

Note on Proposition XX (pp. 24–25). Here Maclaurin is referring to Newton's *Opticks* ([82, 83], see also [87]). First he mentions the bending of rays of light 'near the corners of bodies', which is known nowadays as diffraction; again it is an effect to be attributed to God. Then he gives the law of refraction, which states in modern terminology:

$$\frac{\sin\theta_1}{\sin\theta_2} = \frac{n_2}{n_1},$$

where θ_1 is the angle of incidence, θ_2 is the angle of refraction, and the ray of light is passing from a medium with refractive index n_1 into one with refractive index n_2. MacLaurin appears to attribute this law to Newton, who gives it in Proposition VI of Book I of [82]. However, it is generally known as Snell's law after the Dutch scientist Willebrord Snell (1580–1626), who formulated a version of it about 1621; Newton certainly recognised Snell's contribution in the Scholium to Proposition XCVI in Book I of [81] ([88]). MacLaurin notes that, in going from a rarer to a denser medium $(n_1 < n_2)$, the refracted ray makes a smaller angle with the normal $(\sin\theta_2 < \sin\theta_1)$, while in passing from a denser to a rarer medium $(n_2 < n_1)$ the reverse will occur unless "the angle of incidence is exceedingly large": specifically, if $n_1 n_2^{-1}\sin\theta_1 > 1$, Snell's law cannot be satisfied and the ray of light must be reflected (total internal reflection). The phrase "alternations of easier reflection and transmission" found in the last sentence refers to Newton's explanation for the formation of the interference patterns, now known as Newton's rings. At an interface a ray of light could be reflected or transmitted – in Newton's terminology ([87], Book II, Part III, Definition): *The returns of the disposition of any Ray to be reflected I will call its* Fits of easy Reflexion, *and those of its disposition to be transmitted its* Fits of easy Transmission, *and the space it passes between every return and the next return, the* Interval of its Fits.

Appendix I

I.1. Hand-written note at the beginning of the copy of MacLaurin's MA dissertation in the Babson Collection[23]

Anno 1717 mense Augusto, cum Abraedoniae pro Professione mathematica in Collegio Novo vacante, Ventilationem publicam in omnibus matheseos partibus Aemuli pericularemur per duas continuas Hepdomadas, Hanc mihi dono dedit, Benevolentiae et amicitiae pignus quam postea Lunevillae culti sumus in Academia Equestri Lotharingiae Ducis anno 1724 cujus post obitum suum non fui immemor. Anno 1746 nam cum Synopsis Philosophiae Newtonianae, opus ejus posthumum, collatitia pecuniae ederetur, familiae suae gratia, viginti quatuor Exemplaria inter amicos meos elocavi.

Translation

In the month of August of the year 1717, when we, the contestants for the vacant mathematical Professorship in the New College of Aberdeen, were examined during two uninterrupted weeks in all parts of mathematics, he gave this to me as a gift, a token of good-will and friendship, which we developed afterwards at Luneville in the Academy of the Order of Knights of the Duke of Lorraine in the year 1724. I was not forgetful of him after his death in the year 1746, for when the Synopsis of Newtonian Philosophy, his posthumous work, was published by subscription, for the benefit of his family, I lent out 24 copies among my friends.

---

[23] The inscription is apparently by Walter Bowman, who was considered along with MacLaurin for the Professorship [90]. He is probably the Walter Bowman (1699–1782) recorded in [54] as a tutor and antiquary. The *Synopsis of Newtonian Philosophy* mentioned is [70] ([73]); Bowman does appear on its subscription list. This note is reproduced with permission (see my Preface).

I.2. MacLaurin's Dedication

Viro Reverendo,

Mro. **DANIELI M^cLAURIN,**

Ecclesiæ ad Cellam Finani Pastori fidelissimo,

Patruo suo spectatissimo,

Ob affectum curamque plane parentalem,

Patris charissimi loco semper honorando,

Dissertationem hancce Philosophicam,
studiorum suorum primitias,

In animi grati & perpetuum addicti tesseram,

D. D. C. q.

Colinus McLaurin, *Auctor.*

Translation

To the Reverend,

Mr. **DANIEL M^cLAURIN,**

Most faithful Minister of the Congregation

at the Church of St Finan,

His most esteemed Uncle,

For his love and wholly parental care,

In the place of his dearest Father,

which is always to be honoured,

Colin McLaurin, *the Author*

Gives, devotes and dedicates

This Philosophical Dissertation,
the first fruits of his studies,

As a token of a grateful soul

and of one who is perpetually indebted.

I.3. Facsimile of MacLaurin's Title Page

Dissertatio Philosophica Inauguralis,

DE

Gravitate, aliisque viribus Naturalibus,

QUAM
Cum annexis Corollariis,

Favente summo Numine,

Auctoritate Dignissimi Vice-cancellarii,

Joannis Stirling, V. D. M. SS. Th. Professoris Primarii;

NECNON

Amplissimi Senatus Academici consensu, & Celeberrimæ
FACUTATIS Artium Decreto:

Pro Gradu Magisterii, summisque in Philosophia & Artibus liberalibus
Privilegiis & Honoribus rite ac legitime consequendis,

In Auditorio publico Academiæ *Glasguensis,*

Ad diem Junii   *hora   post meridiem,*

Propugnabit COLINUS MᶜLAURIN, *Scotus.*

Prov. 3. 19. *Deus sapientia fundavit terram, stabilivit Cælos prudentia.*

I.4. MacLaurin's Latin Text

Dissertatio Philosophica,
De Gravitate, aliisque viribus Naturalibus.

I. Inter varia naturae corporeae phaenomena, duo sunt, quae, uti in se spectata prae caeteris fere omnibus maxime sunt insignia, ita omnis aevi Philosophos plurimum exercuerunt. Alterum est, *generalis* illa omnium corporum circa terrae superficiem versantium ad ejus centrum *Tendentia*, quae vulgo *Gravitas* appellatur; alterum,*regularis*, certisque periodis recurrens, *Planetarum*, in orbitis suis *Gyratio*. Ad mechanicam istorum phaenomenon explicationem, hypotheses variae a variis excogitatae sunt. Harum aequuum examen viam struet ad explicandam et adstruendam generalem illam universalis *Gravitatis* legem, ad quam, tanquam commune principium, referendos constabit duos istos effectus nobilissimos; tametsi prima specie nihil inter se commune habere videantur: unde etiam occasionem arripiemus, alias quasdam naturae vires, ad aliorum quorundam phaenomenon solutionem, quae rationibus pariter mechanicis explicare aggressi sunt Philosophi, necessario supponendas, obiter considerandi.

II. Ut a corporum terrestrium gravitate initium fiat, primum examen meretur *Cartesii*, ejusque sectatorum sententia. Illi, inter caeteros, quos *materiae coelesti* adfingunt, stupendos effectus, gravitatem quoque derivant a perenni ejus circa terram gyratione rapidissima; quae violentum a centro motus recedendi conatum isti materiae necessario indit, quo corpora terrestria, vim multo minorem habentia, versus terrae centrum detruduntur. Quemadmodum aqua, vel quodvis aliud fluidum, sibi injectum corpus specifice levius sursum pellit. Caeterum hoc obvio incommodo laborat haec hypothesis, quod *materiae coelesti*, (cujus nulla nobis se produnt in rerum natura vestigia) motum valde rapidum, et quidem circularem, tribuat, cujus ortum et conservationem mechanice explicare, res aeque magni operis ac laboris est, atque ipsius gravitatis rationem reddere. Praeterea, cum haec ipsa materia necessario sit supponenda omnis gravitatis expers, quid tandem ejus conatum centrifugum, usque adeo violentum, cohibere potest? Non alterius fluidi circumambientis pressura; cum necesse sit illud ab hac materia vicissim premi, motumque ei communicari; atque cum hoc fluidum ab alio aliquo superincumbente fluido premi supponendum sit, illud etiam vicissim premet: quo pacto fiet, ut hujus materiae motus in infinitum propagatus perpetuo decrescat, et tandem in nihilum redigatur. Denique, cum haec materia suos gyros necessario perficiat in circulis Aequatori parallelis, necesse erit omnia gravia in istorum circulorum planis descendere, et proinde in lineis non versus terrae centrum tendentibus, sed ejus axi perpendicularibus; omnino contra experientiam.

III. Alii *Gravitatem* ab *Aeris* superincumbentis *pressura* oriri afferunt; non advertentes omnem aeris pressuram ab hac ipsa gravitate dependere: ipsa enim ejus vis elastica, sine aliqua vi elaterium tendente, nullam potest diutinam pressuram efficere, utpote qua cito totus aer multo rarior redderetur, quam in machina pneumatica facile reddi potest; in qua tamen aeris raritatem, plerisque, si non omnibus animalibus, certam perniciem adferre cernimus. Caeterum haec sententia ex eo efficacissime refutatur, quod gravitatis vis multo deprehendatur validior, ubi aeris pressura aufertur, quam ea manente: tantum igitur abest ut illa pressura sit causa gravitatis, ut contra hujus effectum in omnibus corporibus imminuat, et in aliquibus penitus tollat: tantundem enim dati cujusvis corporis gravitati detrahit, quantum aequale est gravitati massae aeris dato corpori mole aequalis; ubi autem residuum est quantitas negativa, uti res se habet in corporibus, quae *Levia* dicuntur, corpora non descendunt sed ascendunt.

IV. Sunt qui *Gravitatem* afferunt esse *Attractionem* ejusdem generis, ac est ea, qua magnes alium magnetem aut ferrum attrahit; et proinde, si haec mechanice explicari possit, (quod plurimi fieri posse existimant) de illa pariter esse philosophandum. Rem autem longe aliter se habere, ostendet breviuscula utriusque generis virium comparatio. Terra, vi gravitatis, corpora quaelibet circumambientia trahit in lineis ad ejus centrum tendentibus, vel accurate, vel quam proxime; idque viribus (ut postea patebit) in paribus a centro distantiis, quantitati materiae in singulis corporibus proportionalibus; in diversis vero intervallis, in duplicata auctarum distantiarum ratione decrescentibus. Magnes, ex adverso, non tam versus centrum, quam alterutrum polorum, attrahit; aequidistantia corpora adeo non attrahit viribus eorum materiae quantitati proportionalibus, ut corporum aequalium alia majori, alia minori, pleraque prorsus nulla vi attrahat: decrescit denique in ratione distantiarum plusquam duplicata.

V. Alia adduci possunt argumenta non pauca, quae evertunt, tum modo propositas, tum alias omnes possibiles hypotheses, mechanicam gravitatis solutionem prae se ferentes; dum nempe evincunt gravium descensum a nullo impulsu corporeo provenire posse. In primis, quandoquidem, ubi velocitates sunt aequales, momenta motus sunt semper ut materiae quantitates; cumque gravia, in eadem a terrae centro distantia, pari velocitate (abstrahendo ab aeris resistentia) versus eam tendant; patet, vires impressas esse directe ut materiae quantitates in ipsis corporibus, nulla figurae, texturae, aut molis habita ratione. Si autem gravitas ab ullo ambientis fluidi impulsu proveniret, ille impulsus vel consisteret in percussione partium fluidi, versus eandem plagam, ad quam urgetur corpus impulsum, libere motarum; vel in pressura totius fluidi, contra impedimentum in altera parte positum validius nitentis: in priore casu, vis imprimeretur pro ratione superficiei, in posteriore, pro ratione molis corporis impulsi; in neutro pro ratione quantitatis materiae. Preaterea omnis impulsus corpus quiescens magis urget, quam corpus in motu positum; ita ut quo majore velocitate moveatur corpus impulsum, eo minus velocitatis incrementum ei addat corpus impellens, donec, corporis impulsi atque impellentis velocitate aequali facta, omnis cesset impulsus, ac motus acceleratio: gravitas autem (ut ex accuratissime institutis experimentis compertum est) corpori celerrime descendenti, et quiescenti, aequali tempore, aequalia addit velocitatis incrementa. Patet igitur gravitatem a nullo impulsu corporeo provenire posse.

VI. Propositioni jam probatae si duae aliae jungantur, patebit, quid de gravitatis causa sit sentiendum. Altera est, quando corpus in quiete positum e loco suo movetur, motum ipsi imprimi ab aliqua causa externa, vel corporea vel incorporea: multo magis, si corpus, versus unam plagam projectum, in plagam directe contrariam retorqueatur, novus ille et contrarius motus a causa externa procedere censendus est. Altera, nullum corpus posse movere aliud, nisi impulsu, *i. e.* corpus nullam vim exerere posse in distans, sive non agere, ubi non est. Sicubi igitur vulgo receptum et concisum loquendi modum sequentes corporum alia corpora *trahentium* aut sine impulsu *repellentium*, faciamus mentionem, indicare volumus istiusmodi phrasibus, non veram et proprie sic dictam motus, de quo agitur, causam, sed occasionem duntaxat, ad cujus praesentiam, secundum generalem aliquam naturae legem, vis ita movendi imprimitur, simulque terminum ad quem, vel a quo, ea vis dirigitur: quod semel monuisse sufficiat. Prior ostendit, corporum terrestrium gravitatem ab aliqua externa causa oriri; posterior, ejus causam non esse rem quamvis corpoream, siquidem *superiore thesi* probatum est, eam non oriri ab impulsu. Quid superest igitur aliud, quam ut gravitatis causa agnoscatur efficax alicujus causae incorporeae et intelligentis voluntas, secundum certam generalem legem, vim suam uniformiter exerentis. Qualis autem sit haec causa intelligens, facile patebit cuivis consideranti, hac ipsa gravitate totam orbis terrae compagem conservari ac firmari; quae alias impetu centrifugo disrupta cito dilaberetur. Gravitas impedit, quo minus montes, maria, urbes, homines, caeteraque animalia, a tellure excussa, per vasta coelorum spatia longe dissipentur. A gravitate pendet tum hominum, tum reliquorum animantium vita et nutritio; ita ut jure meritissimo gravitatis Auctor agnoscendus sit terrae dominus et hominum conservator.

VII. Hujusmodi etiam gravitate, Planetarum reliquorum et *Solis* partes inter se uniri, probant eorum circa axes suos rotationes, necessario producentes conatum centrifugum, partes istas cito disjecturum, ni a gravitate cohiberentur: quae quidem rotationes, in *Sole* et plerisque Planetarum observationibus innotescunt; in *Jove* autem praecipue, non tantum per successivam macularum gyrationem, sed et per figuram sphaeroidicam, ex eadem rotatione oriundam, qui ob corporis magnitudinem et motus rapiditatem satis est sensibilis. Hanc autem partium singulorum Planetarum versus se mutuo gravitatem, in omnibus articulis cum nostra terrestri convenire, ex postea dicendis patebit.

VIII. Sed neque his cancellis continetur hujus principii efficacia; vim enim illam, qua Planetae in orbitis suis retinentur, ejusdem plane esse generis atque illam, qua corpora terrestria versus terrae centrum detruduntur, accurata istorum effectuum collatio satis evidenter ostendet. Pridem demonstratum est, corpus, quod circa alterum ita movetur, ut radiis ad ejus centrum ductis areas describat temporibus proportionales, in orbita sua retineri per vim versus ejus alterius centrum perpetuo urgentem. Cum igitur compertum sit, rem ita se habere in Planetis omnibus primariis, et Cometis respectu *Solis*, secundariis vero respectu suorum primariorum; hinc constat, vi, qua Planetae in orbitis suis curvilineis retinentur, cum corporum terrestrium gravitate, hoc esse commune, quod versus alicujus magni corporis centrum tendant. Earum in caeteris articulis convenientia non minus evidenter probari potest.

IX. Et primo, *Lunae* vim centripetam (qua eam versus *Terrae* centrum urgeri ex modo dictis patet) eandem esse cum gravitate nostra terrestri ita evincitur. Gravitas (secundum accuratissime instituta pendulorum experimenta) corpora terrestria depellit, uno temporis minuto secundo, per pedes *Parisienses* $15\frac{1}{12}$; et proinde (cum spatia gravibus percursa sint ut quadrata temporum) minuto primo per pedes $60 \times 60 \times 15\frac{1}{12}$: quo eodem tempore *Luna* deprehenditur a tangente, versus *Terram* deflecti per longitudinem pedum $15\frac{1}{12}$: tantum enim esse arcus eo tempore descripti sinum versum, temporis periodici et orbitae amplitudinis collatio satis ostendit: vis igitur *Lunae* acceleratrix versus *Terrae* centrum, est ad vim corporum terrestrium acceleratricem versus idem, ut $15\frac{1}{12}$ ad $60 \times 60 \times 15\frac{1}{12}$, sive ut unum ad 60×60. Atque cum *Lunae* distantia mediocris a *Terrae* centro sit corporum terrestrium circa ejus superficiem versantium distantiae ab eodem sexagecupla; patet, corpora terrestria, atque *Lunam*, ad *Terrae* centrum urgeri viribus, quae sunt quadratis distantiarum ab eodem reciproce proportionales. Cum porro haec ipsa sit ratio virium *Lunae* centripetarum, in diversis partibus ejus orbitae, utpote Ellipticae, circa *Terram* in foco positam descriptae, patet corpora terrestria et *Lunam*, eadem vi, secundum dictam legem in diversis distantiis variata, ad *Terrae* centrum urgeri.

X. Praeterea, cum haec eadem lex, nempe ut vires centripetae sint [reciproce] ut quadrata distantiarum, obtineat in omnibus corporibus, sectionem quamvis conicam, circa aliud in foco positum, describentibus; cumque ejusmodi comperiantur esse orbitae omnium Planetarum et Cometarum (si forte *Joviales* excipias, quorum orbitae perfecte circulares, si seorsum spectentur, cum qualibet vis centripetae lege conciliari possunt;) patet eorum omnium vires centripetas ejusdem esse generis ac est ea vis qua *Luna* et corpora terrestria versus *Terrae* centrum urgentur.

XI. Haec eadem virium centripetarum lex, non minus obtinet in diversis Planetis circa idem centrale corpus revolventibus, quam in eodem Planeta in diversis a corpore, versus quod tendit, distantiis: quippe demonstratum est, ubi plura corpora circa idem centrale corpus ita revolvuntur, ut quadrata temporum periodicorum sint in triplicata ratione mediocrium distantiarum, ea omnia ad istud corpus centrale viribus distantiarum quadratis ab eodem reciproce proportionalibus urgeri. Planetas autem omnes, qui circa idem centrale corpus revolvuntur, eam ipsam distantiarum et temporum rationem servare accuratissimis observationibus compertum est.

XII. Cum igitur vis acceleratrix corporum terrestrium versus *Terram*, et Planetarum simul ac Cometarum versus propria sua centralia corpora, decrescat in ratione distantiarum auctarum duplicata; erit haec vis in diversis corporibus, versus idem centrum, in eadem ab eo distantia, tendentibus aequalis; atque adeo eorum vires motrices, sive pondera erunt materiae quantitatibus in iis proportionalia. Cum porro actioni semper aequalis sit reactio, istius corporis centralis versus illa altera tendentia eorum ponderi erit aequalis; atque adeo materiae quantitati in iis proportionalis. Patet igitur universaliter pondera corporum esse in ratione composita ex directis rationibus quantitatum materiae corporum gravitantium, et corporum in quae gravitant, et reciproca quadratorum distantiarum. Cum itaque Planetarum Cometarumque vires centripetae, et corporum terrestrium gravitas ejusdem plane generis sint, nulla est ratio cur non putemus illas aeque ac hanc, ad efficacem Auctoris sapientissimi potentissimique voluntatem uniformiter agentem, tanquam unicam causam, referendas esse.

XIII. Interim hoc phaenomenon, ut alia fere omnia, mechanice solvere aggressi sunt *Cartesiani*: quorum hypotheseos refutatio omnem mechanicae explicationis spem debet perimere. Secundum eos, *Sol* rotando circa axem suum fluidum quoddam subtile, eique innatantes primarios Planetas circumfert; qui singuli vortices quoque suos habent, in quorum nonnullis Secundarii deferuntur. Sed primo, cum Planetae circulos non describant, in vorticibus infinite extensis, aut vase sphaerico inclusis, circumferri non possunt; si autem vorticis limites aliter disponantur, Planetae tanto magis a via circulari deviabunt, quanto longius a centro distant; atque eorum omnium Aphelia in eadem coeli regione reperientur: cum contra Planetarum inferiorum excentricitas longe major sit quam superiorum; *Martis Venerisque* aphelia propemodum opposita sint; eorum enim distantia in principio *Virginis* est fere sesquialtera eorundem distantiae in principio *Piscium*. Quae observatio aliud suppeditat argumentum contra hypothesin vorticosam. Cum enim fluidi per canales inaequaliter amplos circumlati motus, in locis angustioribus citatior esse debeat; patet, secundum hypothesin *Cartesianam*, fluidum cui *Terra* innatat (et proinde ipsam terram) duabus istis orbitis intermedium, in principio *Piscium*, quam *Virginis*, velocius ferri debere: quod observationibus plane repugnat. Adhaec si vortices sint homogenei, tempora periodica erunt ut quadrata distantiarum; sin heterogenei sint, et partes a centro remotiores sint crassiores, ut *Cartesius* voluit, et ratio suadet, tempora periodica erunt ut altiores quaedam distantiarum potentiae; cum tamen Planetarum tempora periodica sint in sesquiplicata tantum mediocrium distantiarum ratione. Efficacissime autem vortices *Cartesianos* refellunt orbitarum planetariarum ad *Solis* axem et ad se invicem inclinatio, et Cometarum motus, nunc Planetarum cursui directe contrarius, nunc eorum orbitis perpendicularis.

XIV. Cum igitur *Materia coelestis* (si ulla sit) cum Planetis non circumferatur, et preaterea eorum motum per tot annorum millia sensibiliter non impediverit, Cometisque per eam velocissime tranantibus adeo facilem aperiat viam; patet, coelorum spatia esse quam liberrima, et proinde materiam nullam in iis reperiri, quae tantorum corporum motui continuo regulariter inflectendo sufficiat. Planetarum igitur ac Cometarum motus in orbitis curvilineis, a nullo imperceptibilium quorumvis corpusculorum impulsu, atque adeo nulla mechanica causa oritur. Atque hinc multum accedit magnificae ideae *thesi 6ta* stabilitae Auctoris gravitatis, quem jam constat, non solum totius terrae, sed etiam coeli dominum, omniumque ejus incolarum Conservatorem esse; qui omnium corporum coelestium compagem conservat; cujus pollenti dextra Planetae, in perpetuos gyros circa commune centrale corpus acti, prohibentur, quo minus impetu centrifugo per vastum inane abrepti, omni, quod a *Sole* jam accipiunt, beneficio privati, perpetuo rigeant frigore, et densissimis tenebris involvantur, atque alia amittant omnia quae ad vegetantium vel animantium conservationem pertinent. Sicut autem impetus hic centrifugus, ni gravitate cohiberetur, omnibus Planetis certam cladem adferret, eas a *Sole* abripiendo, ita non minus certam perniciem iis inferret gravitas, eos in ardentem *Solis* atmosphaeram praecipitando, ni ipsis impressus fuisset motus projectilis: duabus vero istis viribus conjunctis, circa *Solem* in linea aliqua curva ferantur necesse est; quae linea erit circularis, si motus projectilis directio, radio ad *Solem* ducto sit perpendicularis, ejusque vis vi gravitatis aequalis: sin alterutra harum conditionum desit, curva illa erit sectio aliqua conica. Quae hic de Primariis dicuntur respectu *Solis*, de Secundariis item, respectu suorum Primariorum, intelligenda sunt.

XV. Jam probatum est, Planetas Primarios in *Solem*, et Secundarios in suos Primarios, gravitare: cum autem corpus quodvis, quod circa aliud, utcunque motum, areas describit temporibus proportionales, praeter vim versus illud aliud tendentem, urgeatur omni vi acceleratrice, qua illud aliud; hinc patet, Planetas Secundarios, non minus quam Primarios, versus *Solem* graves esse. Caeterum non modo Planetis Primariis cum *Sole*, et Secundariis cum *Sole* et suis Primariis; sed et Planetis etiam ejusdem ordinis, Primariis *ex. gr.* inter se, mutuam gravitationem intercedere, liquet ex quibusdam eorum motuum perturbationibus, a nulla alia causa deducendis; quales sunt *Apsidum Nodorumque* migrationes, *etc.* quae sat sensibiles quandoque se reddunt, praesertim in *Jove* et *Saturno* circa conjunctiones istorum Planetarum heliocentricas, ob vastam eorum magnitudinem, et a *Sole* distantiam, motusque simul tarditatem. Cum porro ipsorum *Satellitum* motus sensibiliter etiam in illis conjunctionibus perturbari deprehendantur, patet,

Planetis Primariis cum aliorum quoque Secundariis gravitationis commercium intercedere.

XVI. Universalis vero illius legis, qua omnia corpora in se mutuo gravitant, efficaciam nihil probat evidentius, quam variae illae, quae omnis aevi Astronomos adeo torferunt, Lunaris motus inaequalitates. Supposita enim gravitationis lege, *Lunae*, cujus distantia a *Terra* adeo magna est (etiam ubi cum *Terrae* distantia a *Sole* confertur) vis acceleratrix versus *Solem*, *Terrae* vi acceleratrice versus eundem aliquando major, aliquando minor, esse debet: quae inequalitas, *Luna* versante in syzygiis, erit maxima; in quadraturis minima, seu nulla; quo fiet ut ejus motus a quadraturis ad syzygias (caeteris paribus) acceleretur, a syzygiis vero ad quadraturas retardetur; atque adeo orbitae ejus curvatura, ac a Primario distantia (caeteris paribus) in his, quam in illis, major erit: unde etiam nec *Luna* semper circa *Terram* areas describit temporibus accurate proportionales: quae omnia cum observatis optime conveniunt. Uno verbo, quaecunque irregularitates in *Lunae* motu observationibus deprehenduntur, (deprehenduntur autem quamplurimae) illae omnes necessaria consequentia a priori deducuntur ex supposita universali quam diximus gravitationis lege; quae igitur plurimum ab iis confirmari existimanda est. Ex eadem etiam lege, ac pari evidentia, deducitur *Aequinoctiorum* nota *Praecessio*, ac *Telluris axis oscillatio*, quae bis quotannis contingit.

XVII. Praeterea, secundum hanc legem, partes fluidi cujusvis terrestris, *Lunae* vel *Soli* directe obversae, notabiliter magis, aversa vero minus, in *Lunam* vel *Solem* gravitant, quam ipsum *Terrae* centrum, sive integra ejus moles complexe sumpta; quo proinde tantum earum versus *Terram* gravitationi detrahitur: partium vero lateralium, seu aversis et obversis intermediarum, versus *Terram* gravitationi, *Solis Lunae*ve attractione cum ea aliquantulum conspirante, nonnihil additur: unde necessario sequitur, dum aversae et obversae leviores, laterales autem graviores sunt, illas ab his sursum premi, donec columnarum altitudine pensent, quod gravitati earum acceleratrici deest: *Solis* autem *Lunae*que vires, fluidorum terrestrium (aeris scil. ac maris) aestum, non duplicem, sed unicum, ex eorum compositione aestimandum, efficiunt; qui, propter diversas istorum Luminarium a *Terra* distantias, et ab Aequatore declinationes, diversus esse debet, idque in triplicata distantiarum istarum ratione reciproca. Atque hinc, nec aliunde, omnia aestus marini Phaenomena facillime deduci possunt; quae igitur jam probato gravitationis principio summam adferunt confirmationem.

XVIII. Praeter illam, de qua hucusque egimus, gravitatem, qua omnes materiae particulae, sine ullo figurarum, formarum, circumstantiarum, aut motuum discrimine, versus se mutuo tendunt, viribus in duplicata distantiarum ratione decrescentibus, est et alia quaedam vis, qua exiguae materiae particulae, se mutuo contingentes, vel contactui proximae, validius quam secundum gravitatis legem modo explicatam, ad se mutuo tendunt: quae vis minuitur in plusquam duplicata ratione auctae distantiae: cumque haec vis, in contactu, vel prope contactum, se tantum exerat, eo validior erit duarum quarumvis materiae particularum cohaesio, quo major sit earum contactus; atque adeo particulae, quae ampliores habent superficies planas, vel saltem sibi mutuo congruentes, firmissime sibi mutuo adhaerent; debilius vero (si omnino) quae sphaericas, aliterve convexas habent superficies: prioris generis particulae, corpus satis durum, posterioris autem, fluidum constituunt; atque ex variis intermediis contactus rationibus, variae oriuntur cohaesiones: uno verbo, hinc, alias insolubilia, *Soliditatis* ac *Fluiditatis* Phaenomena facile solvi possunt. Cum autem haec vis in minimis distantiis aliquanter se exerat, quamvis aliqua externa vi corporis partes nonnihil separentur, si modo cum novis particulis non arctius coalescant; sublata illa externa vi, in priores suos contactus et cohaesiones redibunt; quo pacto corpus pristinam suam figuram recuperabit; quam secus necessario deperdet. Atque hinc *Elasticitatis* et *Mollitiei* natura optime enucleatur. Atque ex his intelligi potest, quam magna, in diversis particulis, ex diversa earum figura et textura, oriatur attractionum diversitas; hinc enim quaedam vix ulla, alia maxima vi, versus se mutuo tendunt: inter haec autem maxime notabilia sunt acida salia, quae in menstruis fere dominantur; ista enim a corporis solvendi particulis attracta, in ejus poros, si modo idoneae sint amplitudinis, adeo magna vi ruunt, ut particulas, ni nimis valide cohaereant, disjungant. Quo pacto omnium corporum solutiones facile explicantur.

XIX. Ex hac etiam materiae minimarum particularum mutua attractione, quam plurima fluidorum phaenomena, alias insolubilia, facile enodari possunt. Ex eo enim, quod aquae particulae, ligni aut vitri particulas validius attrahant quam se invicem, oritur notum istud phaenomenon, quod aqua, vase ligneo aut vitreo inclusa, altior sit prope vasis latera quam alibi; atque adeo in tubis minimis ei aliquantulum immersis altior sit quam extra tubum; cum autem argenti vivi particulae validius se invicem quam ligni aut vitri particulas attrahant, res in illo prorsus contrario modo se habet. Hinc etiam sit, quod aquae, aliorumque fluidorum guttulae, cum gravitatis vi cadere deberent, a vitro, ligno, aliisque plerisque corporibus, suspendantur. Atque ut ex aequali partium in Planetis versus se invicem gravitate, sphaerica Planetarum figura necessario oritur, ita ex aequali particularum aquae, argenti vivi, et similium fluidorum, sibi mutuo admodum approximantium, vi centripeta, oritur guttularum in istis fluidis figura sphaerica. Ex modo dictis etiam intelligi potest ratio congruitatis aquae cum ligno, vitro, aliisque corporibus, et argenti vivi cum iisdem incongruitatis; ac pari facilitate omnia reliqua fluidorum congruitatis ac incongruitatis phaenomena solvuntur. Hinc denique patet, quare arenulae, aliaque corpuscula nonnulla, aqua specifice graviora, in ea tamen non demergantur: mutua scilicet aquae particularum attractio, exiguam licet, aliquam tamen producit tenacitatem, cui superandae par non est istorum corpusculorum gravitas. Horum phaenomenon pleraque a quamplurimis per aeris actionem explicabantur; quorum error ex eo evincitur, quod haec phenomena etiam in vacuo obtinere deprehendantur.

XX. Supposita eadem materiae particularum mutua attractione, fermentationis, chrystallizationis, praecipitationis, fluidorum concretionis, electricitatis phaenomena, aliaque plurima, facillime explicari possunt, quibus immorari non licet. Explicatio autem refractionis radiorum lucis, quae hinc suppetit, nobilior est, quam ut eam prorsus intactam praeterire fas sit. Lucis exiguae particulae sive radii (ut ex egregii *Neutoni* observationibus constat) in transitu suo prope corporum angulos incurvantur, idque eo magis quo propius ad corpora accedant: quam incurvationem adeo regularem a nullo effluviorum impulsu provenire, satis est manifestum; sed ab aliqua vi prorsus amechanica, quae ab Auctore naturae, juxta certam legem, iis imprimitur, pro variis distantiis a corporibus, ad quae appellunt, vel a quibus recedunt: cujusmodi vi supposita demonstravit modo laudatus Auctor, necessario sequi sinum refractionis esse ad sinum incidentiae quaecunque sit incidentiae obliquitas, in data semper ratione; quod ita se habere experientia docet. Necesse est igitur ut radii, qui e medio rariore in aliud densius, aut quacunque ratione magis attractivum, oblique incidunt, ab hoc densiore attracti, prius incurventur quam illud attingant, ita ut linea directionis radii, postquam corpus intraverit, minorem cum perpendiculo faciat angulum, quam ante incurvationem: atque hinc oritur refractio ad perpendiculum. Si autem lucis radius e densiore medio in rarius, aut certe minus attractivum, oblique incidat; ob majorem prioris attractionem, versus id, in ipso exitu, vel statim post exitum, incurvabitur, ita ut nunc majorem angulum cum perpendiculo faciat radii directio, quam antea: atque hinc oritur refractio a perpendiculo. Si autem in hoc casu angulus incidentiae sit valde magnus, refractio in reflexionem mutabitur; ita ut angulus incidentiae sit aequalis angulo reflexionis: patet interim lucis particulae motum in priore casu accelerari, in posteriore autem retardari: atque hinc sit quod lucis velocitas in medio densiori plerumque major sit, quam in rariori. Praeterea cum radius versus partes tantum perpendiculariter sibi subjectas impellatur, patet radium per totum incurvationis tempus versari in eodem plano ad superficiem refringentem perpendiculari. Porro, ex diversis ipsorum lucis radiorum formis, vel forte velocitatibus, diversae inter lucis radios aliaque corpora attractiones, atque adeo diversi refrangibilitatum gradus oriuntur. Ex aliquo etiam simili principio arcessendae sunt stupendae istae facilioris reflexionis et transmissionis vices; quas in luminis radiis obtinere plurimis experimentis idem egregius Auctor demonstravit.

COROLLARIA.

I. *Mentis natura simplex et incomposita non patitur eam in ulla parte spatii ita existere, ut cum ea coextendatur; nec tamen impedit, quo minus uni loco, ei scil. ubi corpus est, ita sit praesens, ut in alio loco similiter praesens non sit.*

II. *Quantumvis reales sive absolutae substantiarum essentiae sint nobis ignotae, haudquaquam inde sequitur, nos de earum affectionibus et mutuis relationibus nihil certi pronunciare posse.*

III. *Philosophia moralis tanquam necessario fundamento innititur summi numinis existentiae et providentiae, praesertim quatenus haec in praemiis poenisque dispensandis se exerit.*

IV. *Quaecunque jura, vitae servandae, vel gravis alicujus damni avertendi causa, alienari possunt, ea per actus, consensum indicantes, licet injustissima ejus in cujus favorem eduntur violentia extortos, eousque alienantur; ut licet hic, vi talis actus, nihil jure petere, aut penes se tenere, queat, ille tamen qui consensu suo aliquod jus ei conferre, atque adeo obligationem sibi contrahere prae se tulit, omnino ex fidelitate teneatur.*

FINIS.

Part II

MacLaurin on Collisions:
Démonstration des Loix du Choc des Corps
(Royal Academy of Sciences, Paris, 1724)

Part II Contents

Introduction to Part II

The prize-essay topics proposed by the Royal Academy of Sciences in Paris for the years 1724 and 1726 were concerned with the collision of bodies: in the competition of 1724[24] the bodies were to be *perfectly hard* and the prize was won by MacLaurin for his essay [66], with which this article is concerned; *elastic bodies* were to be considered for 1726, when the winning essay was that of Père Maziere, described as Prêtre de l'Oratoire [76]. In a Notice prefixed to the published version of MacLaurin's essay (see Appendix II.1, p. 79) it was stated on behalf of the Academy that many of the submissions, while excellent in themselves, had not dealt with the topic as proposed. Amongst these was an extensive work by Jean Bernoulli, which had apparently been submitted in both 1724 and 1726 and had been praised on both occasions. Bernoulli's essay [13] was also published in the volumes containing the prize essays [1]; one reason for this may have been the desire to present both sides of an on-going controversy concerning the force of a moving body (see below). In fact, MacLaurin had also transgressed the limits of the proposed question, dealing not only with the collision of perfectly hard bodies but also with cases of elastic collisions.

The collision of elastic spherical bodies moving uniformly in the same straight line is governed by the equations (see, for example, [102], pp. 276–277)

$$m_1 v_1 + m_2 v_2 = m_1 u_1 + m_2 u_2 \,, \qquad \text{(i)}$$

$$v_2 - v_1 = -e(u_2 - u_1) \,, \qquad \text{(ii)}$$

where the bodies have masses, initial velocities and final velocities m_1, m_2, u_1, u_2, v_1, v_2, respectively, and e is the coefficient of restitution ($0 < e \leq 1$). Equation (i) is the *law of conservation of linear momentum* and (ii) is the *law of restitution*.

For *perfectly hard* (inelastic) bodies we have $e = 0$ in (ii), so that we obtain $v_1 = v_2 = v$ and then from (i)

---

[24]According to Jean Bernoulli ([13], p. 4, §2) the precise title was *Quelles sont les loix suivant lesquelles un corps parfaitement dur, mis en mouvement, en meut un autre de même nature, soit en repos, soit en mouvement, qu'il rencontre, soit dans la vuide, soit dans la plein*, and the prize amounted to 2,500 livres.

$$v = \frac{m_1 u_1 + m_2 u_2}{m_1 + m_2} \, . \tag{iii}$$

When $e = 1$ the bodies are *perfectly elastic*. In this case equation (ii) becomes

$$v_2 - v_1 = u_1 - u_2 \, , \tag{iv}$$

that is to say, the relative velocity of the two spheres has the same magnitude before and after the collision, but the directions are opposite. From (i) and (iv) we obtain

$$v_1 = \frac{m_1 u_1 - m_2 u_1 + 2 m_2 u_2}{m_1 + m_2} \, , \quad v_2 = \frac{m_2 u_2 - m_1 u_2 + 2 m_1 u_1}{m_1 + m_2} \, . \tag{v}$$

A simple calculation now yields

$$m_1 v_1^2 + m_2 v_2^2 = m_1 u_1^2 + m_2 u_2^2 \, , \tag{vi}$$

which is equivalent to the assertion that kinetic energy is conserved. Equations (i), (iv) and (vi) are what Bernoulli refers to as the "three laws which are constantly conserved in the direct collision of two bodies" ([13], Chapitre X).

The above laws were essentially known well before the end of the seventeenth century. For example, Newton gives the law of restitution in the Scholium to the Laws of Motion in Book I of the *Principia* (see pp. 20–23 of [81] ([88])). Newton also refers to the work of Huygens, Wallis and Wren which was published in the *Philosophical Transactions* in 1668 and 1669 ([59, 112, 118]). The Royal Society had become interested in the collision of bodies in the mid-1660s and had invited these three, who were already recognised authorities on the subject, to submit accounts of their discoveries. We note in particular that Wren had dealt with perfectly elastic collisions, his results being in agreement with corresponding results of Huygens, who also gave equation (vi). In fact, Huygens had already completed a treatise on motion and collision in 1656, although it was only published posthumously in 1703 (see [57]); the rules which he presented in [59] were also published about the same time in [58]. Wallis had apparently considered perfectly hard bodies, for which he also touched upon oblique collisions; his celebrated *Mechanica* [113], whose third part deals with collisions, was published during the period 1670–1671.[25]

The study of collisions was at the centre of a major controversy over what should be understood as the *force of a moving body*. According to the Newtonian view, this force is (proportional to) the *quantity of motion*, namely, the product of the body's mass and its velocity. Against this was the Leibniz–Huygens idea that this force is proportional to the product of the mass and

---

[25]An interesting account of these contributions of Huygens, Wallis and Wren to the *Philosophical Transactions* and of related matters will be found in [52]. See also [53], where correspondence concerning these publications can be found. Both references contain a translation of Wren's paper.

the square of the velocity (Leibniz: *vis viva*; Huygens: *vis ascendens*). In the former case we are dealing with *momentum*, a vector quantity, and in the latter we are effectively concerned with the scalar quantity *kinetic energy*.[26] Both, of course, have their separate roles to play in mechanics, but these were not yet understood and the situation may have been confused further by the fact that in the collision of perfectly elastic bodies both momentum and kinetic energy are conserved (equations (i) and (vi)). MacLaurin, as one might expect, was a fervent advocate of the Newtonian view, as was Père Maziere in [76], while Bernoulli followed Leibniz in [13].

In 1722 the Dutch scientist and philosopher Willem Jacob 'sGravesande (1688–1742) published his "Essai d'une nouvelle théorie du choc des corps" [44, 45], in which he founded the study of collisions on the Leibniz–Huygens concept of force. A substantial part of MacLaurin's essay (its Section II) is devoted to an attempt to demolish 'sGravesande's work, starting with its foundation. Although 'sGravesande recovered certain results that had previously been established on Newtonian principles and which appear in MacLaurin's essay, MacLaurin argued that 'sGravesande's derivations are based on a Proposition for which he had given a deficient proof; for his part, 'sGravesande explained the concordance of results as being due to compensating errors in the Newtonian approach.

There was much criticism of 'sGravesande's work in addition to that contained in MacLaurin's essay; in particular, a very public and somewhat scurrilous attack was made in the *Philosophical Transactions* by Samuel Clarke[27] [29] (see also [31], pp. 737–740), who had already inveighed against Leibniz on the same topic (see [11], pp. 121–125). To answer his critics 'sGravesande produced his *Supplément* (1722) [45] and *Remarques* (1730) [46].[28] A direct reply was made to Clarke in the first part of [46]; otherwise, 'sGravesande did not name his critics but concentrated on refuting their criticisms – parts of [46] appear to deal with points from MacLaurin's essay. It is interesting to note that 'sGravesande was in other matters a promoter of Newtonian philosophy. His *Physices elementa mathematica, experimentis confirmata. Sive, introductio ad philosophiam Newtonianem* (Leiden, 1720, 1721) went through many editions and was translated into English and French; in later editions he converted to the Leibniz–Huygens concept of the force of a moving body and

---

[26]The interest appeared to be in comparing forces, so that the constants of proportionality were largely irrelevant. The point is made in [16] that the $\frac{1}{2}$ in the definition of kinetic energy as $\frac{1}{2}mv^2$ only became significant with the later study of work and power.

[27]Among the notes which Clarke supplied for Rohault's *Physica* there is a discussion of collisions of perfectly elastic spherical bodies. See [93], Part I, Chapter 11, pp. 45–48.

[28]Part of the contents of [46] may be much earlier than 1730. Publication of the *Journal littéraire de la Haye*, in which the article appeared, was interrupted in 1722 and only recommenced in 1729.

included much material on collisions (see, for example, the French translation [48], Chapitres IV–VIII in Tome I, Livre II, Partie II).

Reproduced in [49] (see Vol. 1, Part 1, pp. xxxvi–xlv) is a letter dated 31 October 1722 from Jean Bernoulli to 'sGravesande in which Bernoulli thanks him for sending the first edition of his *Physices*, his essay on collisions and the pamphlet [47]. In it Bernoulli chides 'sGravesande for his sycophantic praise of Newton and unleashes an attack on "les Anglois," which begins with the following, translated from the French:[29]

> But the English, whose sentiments it appears you have espoused and have taken sides under their flag, at least in matters of Physics; the English, I say, what will they say when they see that you have fallen into one of the heresies of Mr. Leibnitz? For, among them everything is a heresy which comes originally from this great man; it is a pity for them, that the first discovery of the true estimation of forces has not been made by Mr. Newton, they would not have failed to extract substance from it to glorify the perspicacity of their nation, and a reason to gloat over the blindness of others; whereas at present it is an error, a reverie, a childish absurdity to think along with Mr. Leibnitz that the force of Bodies is proportional to the masses and to the squares of the velocities and that the quantity of forces is therefore quite different from what is commonly called the *Quantity of Motion*.

Thus we see that the matter was pursued with heated passion on both sides, 'sGravesande being something of an innocent victim who appears only to have wanted to present his new theory and discuss it in a courteous manner.

MacLaurin's essay was written during his sojourn in Lorraine.[30] Its first section contains a review of the known and generally accepted laws of motion. As already noted, the second section is devoted to disputing, often in fairly intemperate terms, 'sGravesande's arguments and the Leibniz–Huygens concept of force; it ends with a categorical statement of the Newtonian view. In the third section MacLaurin deals with direct collisions of perfectly hard bodies (Propositions I and II and their Corollaries) and of perfectly elastic bodies (Proposition III and its Corollaries); in Proposition IV he notes the necessary modification when the elasticity is not perfect $(0 < e < 1)$. Also in this section MacLaurin makes a further attack on 'sGravesande, this time for his ideas on perpetual motion [47]. Finally, in the fourth section MacLaurin presents a nice geometrical treatment of oblique collisions, which was possibly

---

[29]Perhaps, as a Scot, MacLaurin was not part of the target of Bernoulli's attack! The letter of course predates MacLaurin's essay. However, he was already known to Bernoulli at this time, for in the same letter Bernoulli mentions that he had received MacLaurin's book, presumably the *Geometria Organica* [65], makes some criticisms of it and, curiously, asks 'sGravesande to thank MacLaurin if the opportunity should arise.

[30]For about two years from 1722 MacLaurin travelled in France as tutor and companion to the son of Lord Polwarth. (See [70] ([73]), pp. iii–iv, and my General Introduction, p. 2.)

original. The essay is given in translation below. This is followed by a series of notes in which aspects of the essay are discussed in detail. I have redrawn MacLaurin's diagrams for Section IV; the originals from [1] are reproduced in Appendix II.2 (p. 80).

MacLaurin continued the study of collisions in his *Treatise of Fluxions* (1742) (see [69], Book I, Chapter XII, Articles 511–520). In particular he obtained there certain results of Huygens, including equation (vi) above, and dealt with multiple collisions in which he extended results contained in Bernoulli's essy [13]; an important method was to study the motion of a system of bodies in terms of that of its centre of gravity. MacLaurin tells us that this material comes from a treatise he had written in 1728 as a supplement to his essay but had not published, although it had been shown to several persons. MacLaurin's reason for not publishing, while uncharacteristic, perhaps puts the whole controversy nicely in context: 'I was unwilling to engage in a dispute that was perplexed by such suppositions, and that after all might seem to be in a great measure about words.' Chapter IV in Book II of [70] ([73]) is devoted to collisions and repeats some of this material.

On pp. 65–76 of [35] Desaguliers reproduces with his own annotations part of a manuscript of MacLaurin's entitled "A Treatise of Motion from Impulse." The relevant part consists of Articles 62–82, which deal with "The Measure of the Force of Bodies in Motion."[31] Here MacLaurin is again concerned to refute the Leibnitz–Huygens idea of the force of a moving body and Desaguliers's notes are aimed at showing that the two theories are equally valid; in his Preface he makes an observation similar to that in the last quotation from MacLaurin above: "... the whole was only a Dispute about Words; the contending Parties meaning different Things by the Word FORCE."

---

[31]I do not know if this manuscript is MacLaurin's 1728 treatise. Desaguliers had been given it by a "Mr Charles, a Mathematician in London," through whom he sought and obtained MacLaurin's approval for its publication. The omitted Articles 1–61 are described as "foreign to our purpose." See also the letter of April 10th 1740 from Desaguliers to MacLaurin ([77], Letter 170, esp. pp. 334–335).

Translation of MacLaurin's Essay

DEMONSTRATION OF THE LAWS OF THE COLLISION OF BODIES.

(p.69) ## SECTION I.

Where we present the Axioms and Principles which are in no way disputed concerning the motion of bodies.

I.

Every body at rest remains in this state until some extraneous cause sets it in motion; and every body in motion continues to move in a straight line, without changing its velocity, as long as no extraneous cause acts on this body.

II.

The change of force, that is to say, its increase or decrease, is always proportional to the applied force, and takes place in the direction of this force.

By *applied force* we mean that which is entirely used up in increasing or decreasing the motion of the body.

III.

The action and the reaction are always equal, and have opposite directions; that is to say, the action and the reaction produce in bodies equal changes of motion.

These three principles are demonstrated by an infinity of experiments. We usually call them *the Laws of motion.*

IV.

The spaces travelled over by two bodies, whose motions are uniform, are always in the ratio composed of those of their velocities and of the times that they are in motion.

V.

The forces of bodies, whose velocities are equal, are proportional to their masses.

VI.

The force produced in a body can never be greater than that which the agent, which transmits its own motion to it, has, if no elasticity enters into their action.

VII.

All the motions, the forces and the collisions of bodies take place in a space which moves forward with a uniform velocity, the same as if this space were absolutely at rest. It is agreed that the motions and the collisions of bodies take place in the same way, whether the Earth turns on its axis, or it is immobile as in the Ptolomeian System. The collisions of bodies on a ship which moves forward with an even motion, are the same as if the ship had no motion.

(p.69)

SECTION II.

Where we show that the forces of bodies are as the products of their masses multiplied by their velocities; and where we examine the opinion of those who claim that the forces are as the masses multiplied by the squares of their velocities.

Since it is absolutely necessary to know how to determine the proportions of forces of bodies in motion before investigating the Laws of their collisions, and since it is disputed that the forces of bodies are as the rectangles or products of their masses by their velocities, it seems to me essential to clarify this matter, and to examine with care the opinion of Mr *Leibnitz*, explained and supported recently in quite a coherent manner by Mr *Sgravezande* in an essay which he has published on the collision of bodies. It is the most fundamental question that one can consider in connection with the collision of bodies; this is why I dwell more particularly on his discussion.

I. Mr *Leibnitz* and Mr *Sgravezande* claim that the forces of bodies are as the products of their masses by the squares of their velocities, and that the forces of equal bodies are as the squares of their velocities. For example, if the velocities of two equal bodies are as 10 and 8, their forces must be as 100 and 64.

Let us suppose therefore that two persons, one on a ship, which moves forward with a uniform motion and a velocity as 2, the other at rest on the shore, throw two equal bodies A and B with equal efforts in the direction of the motion of the ship, and that the body B which was at rest gains a velocity as 8. It is clear by the seventh Principle, that the body A will move forward in the ship with a velocity as 8 also, and in the air with a velocity as 10, the sum of the velocity of the ship and of its relative velocity in the ship. The force of the body A, before it had this increase, was as 4, according to Mr *Leibnitz*, its velocity being as 2. The increase of force which it receives is equal to that of the body B by the seventh Principle, that is to say, to 64: thus its total force will be as $64 + 4 = 68$. But because the velocity is as 10, its force must be as 100, and these two forces are contradictory. Therefore their forces cannot be as the squares of their velocities.

If we suppose that another body C equal to the bodies A and B is thrown in the same direction and with the same effort on a ship which moves forward with a velocity as 4, the total velocity of the body C will be as 12, and its force in the air as $12 \times 12 = 144$. Taking away from this 16, which was its force before the increase which it has received, the residue 128 is the force added to the body C by the same effort which had added 96 degrees of force to the body A, and 64 to the body B, according to the system of Mr *Leibnitz*. However, it is clear that these increases must be equal, just as much by the second as by the seventh Principle.

Again to give greater clarity to this reasoning, let us suppose that the two bodies A and B come to strike against some invincible objects positioned one on the ship, the other on the shore, and that the bodies have no elasticity; it is clear that they will lose equal quantities of force, and that the collisions will be the same by the seventh Principle. But the body B will lose 64 degrees of force, which is everything that it had received. In losing 64, the body A will therefore have the residue $100 - 64 = 36$. But, as A loses all its velocity, apart from the two degrees which it had in common with the ship from the beginning, there only remain four degrees of force with it; and these two forces are again contradictory.

Finally, if the system of these authors were true, the motions and the collisions of bodies contained in a space which moves forward uniformly, would be quite different from the motions and collisions of the same bodies when the space remains at rest. In their system it would always have been easy to distinguish relative motions from absolute motions, a task which has been regarded on several occasions as one of the most difficult in Physics.

We extract a similar argument from the motion of elastic bodies. Let there be two equal elastic bodies A and B, which go in the same direction with velocities as 10 and 5; it is known that if they had no elasticity, they would have after their collision a common velocity as $7\frac{1}{2}$: but if they are perfectly elastic, they will change their velocities, and the body A will have 5 and B 10 degrees of velocity. Mr *Sgravezande* agrees in his Prop. 25 that elasticity acts on the bodies as if they were at rest: and because elasticity separates them with 5 degrees of velocity, it is necessary that it applies $2\frac{1}{2}$ degrees of velocity to each body, that is to say, $\frac{25}{4}$ degrees of force. Without the action of elasticity the force of the body A would have been the square of $7\frac{1}{2}$, that is to say, $\frac{225}{4}$; elasticity takes away $\frac{25}{4}$ degrees from it: therefore it must retain $\frac{225-25}{4}$ degrees of force, that is to say, 50 degrees; but since its velocity is only 5, its force can only be 25. These two forces are contradictory: from this we have to conclude that it is impossible to reconcile their principle with experiments.

We could expand further on the arguments which could be extracted from the motions of elastic bodies, but let us pass rather to that which proves more directly that the force is as the mass multiplied by the velocity.

II. It is agreed that two bodies, whose velocities are in the inverse ratio of their masses, and whose directions are opposite, remain at rest after collision. Mr *Sgravezande* acknowledges this. We find that two bodies A and B, which are as 3 and 1 with velocities as 1 and 3, remain at rest after their collision, if they have no elasticity. Their forces, according to Mr *Sgravezande*, are as 9 to 3, or 3 to 1: but according to us their forces are as 3 to 3, or 1 to 1; that is to say, they are equal. We have formerly regarded this experiment as proof that the forces were as the velocities, and not as their squares multiplied by the masses. We have believed that the forces of bodies which destroy each other, must be equal, and as a consequence, that the forces were as the masses multiplied by the velocities. In the other system it is necessary that one force stops another force, of which it has only the third part, or even in other examples, one force must stop an opposite force, of which it is only the thousandth or ten-thousandth part. It is claimed that the larger force loses all its advantage in breaking down the parts[32] of the other. But this answer does not remove the difficulty; it is said that these forces do not destroy each other, but that they are used up in breaking down their parts mutually. Now since these actions are mutual and opposite, and since they begin and end at the same time, and since they continue without prevailing one on the other while they act, I do not understand how they can produce such different effects, one losing sometimes a thousand, or even ten-thousand times more than the other.

In the system of Mr *Sgravezande* it would have been much more natural to believe that, on meeting each other, two bodies, which are as 9 and 1 with velocities as 1 and 3 and have their masses in the inverse ratio of the squares of their velocities and consequently their forces equal, would always have to act with equal and opposite forces in order to break down their parts mutually, and as a consequence would always have to lose equal forces and both remain finally at rest; this is extremely contrary to experiment. To resolve these difficulties, he is obliged to assert that, when two bodies meet each other with velocities which are in the inverse ratio of their masses, the big body resists the other, not only by its force, but also by its inertia; this I regard as a tacit admission that the two forces of the bodies are effectively

---

[32]The French phrase here is 'en enfonçant les parties'; other parts of the verb *enfoncer* are encountered in the essay and have been translated in a corresponding way.

equal in this case: and I find that the author balances thereby the too large force which he had given to the small body as a result of its velocity. In the collisions of these two bodies all the resistance which the big body makes, whatever it may be (and which is equal to the force which is used up in the small body, according to the admission of the author) must diminish equally the forces of the two bodies. Therefore, since in his system the force of the big body is much smaller, it must be used up before the other: the latter, finding no more resistance, must take away both bodies. That seems to me to be an incontestable consequence of our third Principle, that the action and the reaction are equal. To grant to the author his arguments on inertia and the resistance of bodies, it would be necessary to change entirely our ideas of force, inertia and motion, and to abandon what is clear enough in order to adopt things which are very deeply obscure.

But if it is surprising that in his system a lesser force can stop a much larger one, it seems even more extraordinary that a force which is only a thousandth part of another can prevail and take it away on this other one. The author answers that the larger force is used up in breaking down the parts of the other body, which is the larger. But it is more natural to believe that the force which supports the opposite action of the other and takes it away again on itself at the end, is the larger, than to believe that it is only its thousandth part.

III. Mr *Sgravezande* claims to deduce from his principle the same Laws for the collisions of bodies which had already been found by our principle and by experiment. His fourteenth Proposition is the basis of all those which follow, and does not seem to have been sufficiently established. He asserts that "the force lost in the collisions of two nonelastic bodies is the same whatever the absolute velocities of these two bodies may be, provided their relative velocity is the same." We will see first of all that the demonstration which he gives of it is not sufficient to establish one of the principal differences of the two systems. He says: "The motion of the two bodies is composed of their common motion and their relative motion. It is clear that the first, in whatever manner it may be changed, cannot change the action of one body on the other; thus this action is always the same as long as the relative velocity does not change. It is on this action or effort of the bodies, the one on[33] the other, that the flattening or breaking down of the parts depends, which consequently will be the same, if the relative velocity is the same." We could believe, from the manner in which he treats this Proposition, that it had been granted in both systems. However, it is very definitely false in the usual system. It is clear by his nineteenth Proposition[34] that he is speaking of the loss of the sum of the absolute forces of the two bodies, and not of that of the sum of their motions in one direction. It is also established that in the usual system the absolute motion which is lost in the collision of two nonelastic bodies, whose

---

[33]Here MacLaurin quotes "sur" whereas "contre" in found in the version in [49].
[34]See Note on Sections I and II, p. 70, for its statement.

directions are opposite, is double the force of that body, which has the least. This must therefore change if the relative velocity remains the same, when the smaller force changes, and cannot change, although the relative velocity becomes greater, if the smaller force remains the same. Let us suppose that two bodies A and B have velocities V and u, and that the sum of their absolute forces before the collision was $AV + Bu$; if the force of the body A was the larger, and if they go in opposite directions, their force after their collision will be $AV - Bu$, and the difference of these forces, or the force lost, will be $AV + Bu - AV + Bu = 2Bu$, that is to say, equal to double the smaller force. The author had said that the forces never destroy each other, but that they use themselves up in breaking down the parts of the bodies which are opposed to them, and which support themselves by their opposite forces. We could extract from this that a force cannot lose much in breaking down the parts of a body, if this body is not supported by an opposite force, or some other obstacle. At least it seems reasonable to believe that the force lost as a result of the collision of bodies which meet each other with opposite directions, must be greater than when one of the two, with a velocity equal to the sum of their velocities, falls on the other at rest; and neverthless the relative velocity is equal in these two cases. It is certain that if the relative velocity is unchanged, the forces of the bodies can change themselves, and as a consequence the resistances which they will make in their collision, the one against the other, their motions being opposite; it follows from this that the breaking down of the parts, and the force lost can vary. If we find that this Proposition is badly founded, we will upset his whole system: for without this, he would never have reconciled his principle with the Laws of collision established by experiment.

Mr *Sgravezande* tries to avoid the force of the experiment of two bodies, whose velocities are in the inverse ratio of their masses, which remain at rest after their collision, claiming that the forces lost by the breaking down of the parts are unequal. But it is certain that two bodies of unequal masses which pull each other with the same force (like two boats which pull each other with the same rope) move forward with velocities which are in the inverse ratio of their masses; and in this case it cannot be claimed that there is any breaking down of the parts, for the bodies do not touch each other. We could extract more and more arguments against his principle, from what has been demonstrated about centrifugal forces, which always balance, when the accelerating forces are in the inverse ratio of the masses of the bodies, about centres of gravity and about the percussion of bodies; but that would lead us too far off. We content ourselves with explaining those which are easiest.

IV. Finally, it is time to examine the reasoning and the experiments, by which the author claims to establish his principle. He is correct in saying "that it requires less effort to give a certain degree of velocity to a body, than to increase by the same degree the velocity of an equal body, but in motion."[35] But it is also true that the effort in the second case does not exercise completely, and does not lose more than in the first: it is clear from this that there is no more increase of force in the second case than in the first. Let us imagine two men A and B each holding a ball, A being at rest while B is on a boat which is in motion: on throwing these balls with equal efforts, the two men add equal velocities to them, if the balls are equal. It is true that B is transported in the boat; but we see that the force with which he is transported is not diminished, and that it has no effect on the ball which he throws. On applying this reasoning to springs, we will find that the author has not succeeded in the demonstration which he gives of the eighth Proposition.[36] It has to be denied that the effort of the springs, which he uses to set the body in motion, is completely employed in moving the body; there is a part which is employed to transport the springs with the velocity which the body has already acquired. That is incontestable; and I am astonished that the author adds at the end of this demonstration that he has disregarded the inertia of the springs themselves. This comes after he had supposed that an infinite number of springs relax in order to give to the last a velocity equal to that which the body had already acquired.

As for the experiments from which he claims to deduce his principle, it suffices to say that the penetrations of bodies in clay are measures which are not sufficiently exact and geometrical for the determination of their forces. It is impossible or very difficult to reduce to an exact calculation the decelerations of a body which falls on such ground. The author acknowledges that only the weight of a body which has no force can sink it into this clay. From this we see that the penetrations are not proportional to the forces, and that when the former are equal, it does not follow that the latter are also equal. It could indeed be useful to investigate how it might come about that the penetrations are equal, the masses of the bodies being in the inverse ratio of the squares of their velocities. But that experiment does not suffice to establish a principle which cannot be reconciled with other incontestable experiments, as we have demonstrated. Finally, after what we have just said, it can be established for the eighth Principle that

VIII.

The forces of bodies are as their masses multiplied by their velocities.

---

[35]See Appendix II.3, p. 81.
[36]See Note on Sections I and II, p. 70, for its statement.

SECTION III.

Where we give the Laws of Direct Collision.

DEFINITION I.

We call the collision of bodies *direct* when their centres of gravity always run along the same straight line, which passes through the spot where they are going to collide, and furthermore is perpendicular to the parts of the surfaces which collide.

DEFINITION II.

We call those bodies whose parts do not yield at all in the collision *perfectly hard.*

DEFINITION III.

We call a body *elastic* when its parts yield in the collision, but afterwards recover their original positions. If they recover with a force equal to that by which they have been broken down, the body is *perfectly elastic.*

DEFINITION IV.

When the parts of a body yield without restoring themselves, we call it *soft.*

Neither perfectly hard nor perfectly elastic bodies are to be found; but that does not prevent us from considering them in Physics. There is no such thing as a mathematical fluid; but that does not prevent us from investigating the properties of such a fluid, and the resistances which it could cause to the motions of bodies. We will begin with hard bodies without elasticity.

PROPOSITION I.

If two perfectly hard bodies go in the same direction, it is necessary to divide the sum of their forces before the collision by the sum of their masses to obtain their common velocity after the collision.

Everything which one of these bodies loses as a result of the collision, the other gains; therefore the sum of their forces after the collision will be the same as the sum of their forces before the collision. Since the bodies have no elasticity, they will not separate after the collision, but will continue their motion both in the same direction, as if they made only one mass with a common velocity. It is clear from this that in order to have this common velocity, it is necessary by the eighth Principle to divide the sum of their forces by the sum of the masses of the two bodies.

COROLLARY I.

If the two bodies are A and B and their velocities V and u, the sum of their forces before the collision must be $AV + Bu$ by the eighth Principle; therefore their common velocity after the collision will be $\frac{AV+Bu}{A+B}$. The force of the body A after the collision will therefore be $\frac{AAV+ABu}{A+B}$, and the force of the body B after the collision will be $\frac{BAV+BBu}{A+B}$.

COROLLARY II.

The force which one of the bodies gains and the other loses is the force produced from $\frac{AB}{A+B}$ multiplied by the difference of the velocities of the two bodies. If V is larger than u, then the body A will lose the force $\frac{AB}{A+B} \times \overline{V-u}$. For, since its force before the collision is AV and its force after the collision is $\frac{AAV+ABu}{A+B}$, their difference $AV - \overline{\frac{AAV+ABu}{A+B}} = \frac{ABV-ABu}{A+B} = \frac{AB}{A+B} \times \overline{V-u}$ gives the force which the body A loses as a result of the collision; this is equal to the force which the body B gains.

PROPOSITION II.

If the motions of the two bodies have opposite directions, the difference of their forces before the collision has to be divided by the sum of their masses, in order to obtain their common velocity after the collision.

After the collision the two bodies move together in the same direction; as a consequence the larger force destroys the smaller, and, in destroying it, it is itself reduced by a quantity equal to this small force by the third Principle. What remains is the difference of the two forces: thus the sum of the forces of the bodies after the collision is just the difference of the forces which they had before the collision. Therefore this difference has to be divided by the sum of the masses of the bodies in order to obtain their common velocity after the collision.

COROLLARY I.

Let us suppose that the body A has the larger force; the common velocity after the collision of the bodies A and B, whose velocities were V and u, will be $\frac{AV-Bu}{A+B}$. The force of the body A will be $\frac{AAV-ABu}{A+B}$, and the force of B will be $\frac{ABV-BBu}{A+B}$.

COROLLARY II.

The force which the body A loses is $AV - \overline{\frac{AAV-ABu}{A+B}} = \frac{AB}{A+B} \times \overline{V+u}$. The force which the body B gains in the direction towards which both go after the collision, is that which the body A loses, and these forces are the same when the relative velocity $V + u$ does not change, because $\frac{AB}{A+B} \times \overline{V+u}$ only changes with $V + u$; but if we speak of losses of absolute forces, the body B loses the difference of Bu and $\frac{ABV-BBu}{A+B}$, that is to say, $\frac{2BBu-ABV+ABu}{A+B}$; if to this we add the force lost by the body A, which is $\frac{ABV+ABu}{A+B}$, the sum $2Bu$ gives the force lost as a result of the collision of the bodies A and B, just as we have calculated it in the sixteenth article above:[37] this quantity changes in proportion to the force of the body B.

PROPOSITION III.

The action of elasticity in the collision of perfectly elastic bodies doubles the changes of the forces which would be produced in the bodies, if they had no elasticity.

The parts of elastic bodies are broken down by the collision, and always yield until the two bodies move forward with a common velocity, as if there had been no elasticity; the relative velocity, which compressed their spring, no longer acting, they relax, and restoring themselves by the same degrees, and with the same forces by which they had been broken down, they produce the same effects, separating the bodies with a relative velocity, equal to that with which they approached before the collision. There is therefore a double increase produced in the force of the body which gains as a result of the collision, and a double decrease in the force of that body which loses as a result of the collision.

COROLLARY I.

Let A and B be two bodies which go in the same direction with velocities V and u; and let B be the body which goes in front. By Corol. 2 of Prop. I the change of force of each body would have been $\frac{AB}{A+B} \times \overline{V-u}$. It is therefore necessary to add $\frac{2AB}{A+B} \times \overline{V-u}$ to the motion of B before the collision, to obtain its motion after the collision; and it is necessary to take away the same amount from the motion of the body A before the collision, to obtain its force after the collision. Therefore the force of B after the collision will be $\frac{BBu+2ABV-ABu}{A+B}$, and its velocity $\frac{Bu+2AV-Au}{A+B}$. The force of the body A will be $\frac{AAV-ABV+2ABu}{A+B}$, and its velocity $\frac{AV-BV+2Bu}{A+B}$.

COROLLARY II.

If the bodies have opposite directions, it is necessary to take away again from the force of the body A in Corol. 1 of Prop. 2 what it has lost $\frac{AB}{A+B} \times \overline{V+u}$, and we will find its force after the collision to be $\frac{AAV-2ABu-ABV}{A+B}$. But it is necessary to add the same amount to the force of the body B, which will therefore be after the collision $\frac{2ABV-BBu+ABu}{A+B}$, and its velocity will be $\frac{2AV-Bu+Au}{A+B}$. The velocity of the body A after the collision is $\frac{AV-2Bu-BV}{A+B}$; and when this expression becomes negative, the body A is knocked back towards the opposite direction.

---

[37] In the original there is a marginal numbering of paragraphs or groups of paragraphs. The number 16 indicates the first paragraph in Article III of Section II.

COROLLARY III.

If the body A strikes a larger body B at rest, this body B will have more force after the collision, than the body A had before the collision. The force of the body B will be $\frac{2ABV}{A+B}$ supposing that V is the velocity of the body A before the collision: but it is clear that, since B is larger than A, the quantity $\frac{2ABV}{A+B}$ exceeds AV by the difference $\frac{AV}{A+B} \times \overline{B-A}$. If the body B strikes another larger body C at rest, the force of C will exceed that of B: and we find by calculation, the details of which we cannot give here, that if eleven elastic bodies in geometric progression of one to ten, were to strike one after the other, the last would have 394 times more force than the smallest had. A very learned author has recently based on this a proof of the possibility of perpetual motion* in the system which puts forces proportional to masses multiplied by velocities, imagining that we could indeed use these 394 degrees of force to give one of them to the first body, and beyond that to make some machine, "for which we see easily," he says, "that the motion would be continued in perpetuity, if the materials do not wear away." But we can only be extremely astonished that the author does not remember that the other ten bodies are knocked back in the opposite direction with 393 degrees of force, and that the sum of all the forces, taking it in one direction, is only one degree; this upsets entirely his reasoning. In this Corol. B gains the force $\frac{AV}{A+B} \times \overline{B-A}$; but the body A is knocked back towards the opposite direction with the same force: therefore the sum of the forces in one direction remains always AV, as it was before the collision.

PROPOSITION IV.

To find the forces after the collision of bodies which are not perfectly elastic, it is necessary to diminish the relative velocity with which they separate after the collision in the ratio of the elastic force.

In the collisions of perfectly elastic bodies, the relative velocity after the collision is equal to the relative velocity before the collision: in the case of less elastic bodies it is less in the proportion by which the effort of the spring which produces the relative velocity after the collision is less strong. The celebrated Mr *Newton* testifies that he has found this principle to conform with experiment. See his *Scholium* on the Laws of motion, in Book I of his *Principia*. He found, for example, that two glass spheres always separate after the collision with a relative velocity, which was to the relative velocity of their meeting, as 15 to 16 approximately, and that the proportion between these relative velocities was constant in bodies of the same nature, as long as the collision did not disturb the parts of the body, in such a way that they could not restore themselves to their initial positions. It follows from this

---

* *See the remarks on the possibility of perpetual motion by Mr Sgravezande.* (MacLaurin's marginal note.)

observation that the velocity of the body A after the collision in the case of Corol. 2 of Prop. 3 must be $\frac{16AV-31Bu-15BV}{16A+16B}$ if we suppose that this body is a glass ball. We could argue in the same way about other bodies, when their elastic force has been determined by experiments.

(p.74)

SECTION IV.

Concerning indirect collision.

Problem.

The directions, the velocities and the diameters of two spherical bodies being given along with their position in some instant before the collision, to find the place where they will meet.

(Fig. 1.) Let the two bodies[38] be A, B; and let us suppose that they start off at the same time from the places marked A and B in the directions AC, BC, and that the velocity of the body A is to the velocity of the body B as AC is to BD. Draw the parallelogram ABHC and draw DH. With centre C and radius equal to the sum of the semidiameters of the two bodies A and B, draw a circular arc which cuts the straight line DH in L and l; draw LN parallel to CA, and NR parallel to CL. I say that the centres of the two bodies will arrive at the same time at the points N and R, and that it is then that the bodies will meet; for DN is to NL or CR, as DB is to BH or AC; and by division BN is to AR as BD is to AC, or as the velocity of the body B is to the velocity of the body A. These spaces BN and AR will therefore be travelled over in the same time, and the centres of the bodies will arrive at the same time at the points N and R; now since by assumption NR is equal to CL, the sum of the semidiameters of the two bodies, the two bodies must touch and collide at that time.

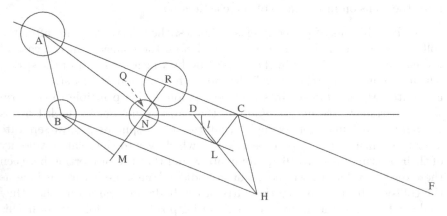

Fig. 1.

---

[38]In the original A, B are used to denote both the bodies and their positions. I have used A, B for the former and A, B for the latter.

COROLLARY I.

The circle drawn with centre C and radius CL, cuts the straight line DH in two points L and l; but when the bodies come to meet from the sides marked A and B, the intersection l is of no use. If the body A comes from the opposite side F, CF and CA being equal, and if the bodies leave from the points F and B together, then in this case to find where they meet, it would be necessary to use the other intersection l, to obtain the position of the bodies in the collision.

COROLLARY II.

If the straight line DH does not enter inside the circle Ll, there will be no point of collision; if the straight line DH touches the circle, the bodies will touch in passing, but there will be no point of collision. If the sine of the angle CDL is not less than the sum of the semidiameters of the bodies A and B, taking DC as radius, there will be no point of collision.

PROPOSITION V.

(Fig. 1.) Let BM, AQ be perpendicular to NR; the actions of the bodies one on the other will be the same as if the body A with a velocity as RQ were to meet the body B with a velocity as MN in the straight line NR.

The velocities of the bodies A, B are proportional to the straight lines AR, BN, and can be represented by these lines. It is known that a force as AR can be resolved into two forces AQ and RQ, and a force as BN into two forces BM and MN. The forces as AQ and BM, having parallel directions and acting in the direction of the tangent of the two bodies, have no effect at all in the collision. Therefore the two bodies will act one on the other as if they were meeting in the direction NR with velocities as RQ and MN.

COROLLARY.

(Fig. 2.) It follows from this Proposition that in order to determine their motions after the collision, it is necessary to suppose that the collision is direct, and that the bodies A and B meet with velocities as QR and MN, and we will find by the Propositions of the previous Section their velocities after the collision in this same direction. Let us suppose that the velocity of the body A after the collision must be Rg, and the velocity of the body B equal to Nm; let Rq be equal and parallel to AQ, and Nl equal and parallel to BM: let the parallelograms Rqag, Nlbm be drawn, and the bodies A and B will continue their motion after the collision in the diagonals Ra and Nb of these parallelograms with velocities as Ra and Nb. It is unnecessary to explain all the special cases of indirect collision; it is easy to apply this general construction always.

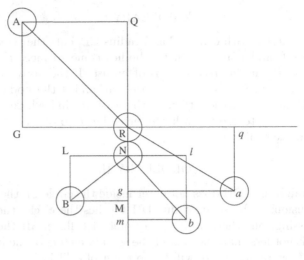

Fig. 2.

There are the Principles and the fundamental Laws of the collision of bodies. In order to explain the more complicated cases of the collisions of irregular bodies, it would be necessary to enter into a long account of deep Geometry. But it is enough to have established the most essential principles which will be able to serve as a foundation for those who wish to push their researches further.

Ac veteres quidem Philosophi in Beatorum Insulis fingunt, qualis natura sit vita Sapientium, quos cura omni liberatos ... nihil aliud esse acturos putant, nisi ut omne tempus in quaerendo, ac discendo, in naturae cognitione consumant. Cic. de fin. lib. V.[39]

THE END.

---

[39]This quotation comes from Cicero's *De Finibus Bonorum et Malorum*, Book V, xix. It has been translated as, "The old philosophers picture what the life of the Wise will be in the Islands of the Blest, and think that being released from all anxiety and needing none of the necessary equipment or accessories of life, they will do nothing but spend their whole time upon study and research in the science of nature" (see [91], pp. 452–455).

Notes on Part II

Note on Sections I and II (pp. 55–61). MacLaurin's seventh principle (Section I) is concerned with inertial frames of reference. He attempts to show in Article I of Section II that the Leibniz–Huygens concept of the force of a moving body is not compatible with this principle, essentially on the grounds that

$$m(v_1 + v_2)^2 \neq mv_1^2 + mv_2^2.$$

MacLaurin's example of the two men throwing objects, one on a ship, the other on the shore, was taken up by 'sGravesande in his *Remarques* [46].[40] In answer to this "*difficulté spécieuse*" 'sGravesande effectively pointed out that, as a result of the reaction produced by the act of throwing, the ship did not remain in its uniform motion. He replaced the passenger by a spring fixed on the ship, which would propel the body, and after making various assertions about the behaviour of the spring he attempted to show that the "force" of the body is indeed 100 and not 68 as MacLaurin claimed. The details of 'sGravesande's analysis are given in translation in Appendix II.4 (pp. 81–83).[41]

In the last example of Article I, which involves elastic bodies, the final velocities stated by MacLaurin follow from equations (iii) and (v) (Introduction, p. 50). He refers to 'sGravesande's Proposition XXV, which states:

> The relative velocity with which two elastic bodies separate after collision is equal to that with which they approached each other.

Here the elasticity is assumed to be perfect.

Referring to equation (iii) of the Introduction, if $m_1 u_1 = -m_2 u_2$, so that the velocities have opposite directions and their magnitudes are inversely proportional to the masses, we have 0 for the common velocity v after collision in the case of perfectly hard bodies, that is to say, the bodies remain at rest after collision. This is what MacLaurin asserts at the beginning of Article II in Section II and proceeds to use as further grounds for denying 'sGravesande's arguments.

---

[40]In fact 'sGravesande had already employed these two men for another purpose in [44]. See Appendix II.3, p. 81.

[41]An almost word-for-word quotation of the second paragraph of MacLaurin's Article I is omitted from that Appendix.

In Article III MacLaurin is concerned with 'sGravesande's Proposition XIV, which he states along with most of 'sGravesande's proof (see Appendix II.5 (p. 84) for the whole proof). MacLaurin seems to be arguing that the Proposition is false if force means momentum. However, although he criticises s'Gravesande's proof, the Proposition is correct for perfectly hard bodies under the Leibniz–Huygens concept of force, which 'sGravesande intended; in this case, according to equation (iii), the loss of "force" is

$$m_1 v_1^2 + m_2 v_2^2 - m_1 u_1^2 - m_2 u_2^2$$
$$= (m_1 + m_2)\left(\frac{m_1 u_1 + m_2 u_2}{m_1 + m_2}\right)^2 - m_1 u_1^2 - m_2 u_2^2$$
$$= \frac{(m_1 u_1 + m_2 u_2)^2 - m_1^2 u_1^2 - m_2^2 u_2^2 - m_1 m_2(u_1^2 + u_2^2)}{m_1 + m_2}$$
$$= -\frac{m_1 m_2(u_1 - u_2)^2}{m_1 + m_2}. \tag{2.1}$$

MacLaurin also refers to 'sGravesande's Proposition XIX, which asserts:

The force lost in the collision of two bodies is proportional to the square of the relative velocity multiplied by the product of the masses divided by the sum of the same masses.

This is just equation (2.1)! In the last paragraph of Article III MacLaurin criticises 'sGravesande's treatment of the case where $m_1 u_1 = -m_2 u_2$ and, consequently, the bodies are at rest after the collision (see above). In this case both sides of (2.1) reduce to $-m_1 u_1^2 - m_2 u_2^2$, that is to say, all energy is lost in the collision.

The statement of 'sGravesande's Proposition VIII is as follows:

In equal bodies the forces are proportional to the squares of their velocities.

This is the fundamental point of contention and MacLaurin attempts to refute it in his Article IV. After the statement of his Proposition VIII 'sGravesande adds, "As this proposition is contested, I shall prove it first by experiment before giving the demonstration." He then proceeds to describe the results of dropping copper balls onto clay, which were apparently compatible with Proposition VIII – certainly, the indentations would depend on energy rather than momentum. MacLaurin's objections on the grounds of exactness seem to be unfounded. In the *Supplément* [45] 'sGravesande describes a further experiment where ivory cylinders with hemispherical ends were dropped onto a moistened marble surface and the imprints measured.[42]

---

[42]In [44] 'sGravesande refers to the use of an apparatus of Mariotte's for his experiments and in [46] he cites Mariotte's treatise on collisions [74]. This work was apparently quite influential. However, Huygens claimed that Mariotte had taken the theory from him and in the late nineteenth century some further doubt was cast on Mariotte's originality by P. G. Tait in [101] following his investiga-

Unlike 'sGravesande, Bernoulli and Maziere, MacLaurin appears not to have been bothered about the mechanism of elasticity. References to the "breaking down of the parts" and to springs relate to the idea that the colliding surfaces are made up of tiny springs, some of which compress on collision and relax as the bodies separate. It is interesting to note that as well as the term *élastique* 'sGravesande uses equivalently *flexible à ressort* (*ressort* ≡ spring, elasticity).

Note on Section III (pp. 62–66). MacLaurin's definitions of *perfectly hard, elastic, perfectly elastic* and *soft* bodies are quite clear. For 'sGravesande, however, perfectly hard bodies, which he did not define formally, were to be rejected on the grounds that none are known; on the other hand, he did study the collision of perfectly elastic bodies, which was surely just as contentious. His Article VII, containing Propositions XII–XXII, is entitled, "Concerning the collision of bodies which are neither perfectly hard nor elastic." Whatever the physical distinction between such bodies and MacLaurin's perfectly hard bodies, they appear to obey the same laws of collision.

As noted in the Introduction (equation (iii), p. 50) the common velocity of two perfectly hard bodies following direct collision is

$$v = \frac{m_1 u_1 + m_2 u_2}{m_1 + m_2}, \tag{3.1}$$

where the bodies have masses, initial velocities and final velocities m_1, m_2, u_1, u_2, v_1, v_2, respectively. This is the content of Propositions I and II. In the first of these, u_1 and u_2 have the same sign, while in the second they have opposite signs; MacLaurin's u and V are magnitudes (speeds) which have to be added or subtracted according as the directions are the same or opposite. The corollaries are concerned with momentum and changes of momentum:

$$m_1 v_1 - m_1 u_1 = m_1 v - m_1 u_1 = \frac{m_1(m_1 u_1 + m_2 u_2) - m_1(m_1 + m_2)u_1}{m_1 + m_2}$$

$$= \frac{m_1 m_2 (u_2 - u_1)}{m_1 + m_2}; \tag{3.2}$$

$$m_2 v_2 - m_2 u_2 = m_2 v - m_2 u_2 = \frac{m_1 m_2 (u_1 - u_2)}{m_1 + m_2}. \tag{3.3}$$

In the final part of Corollary II of Proposition II, MacLaurin deals with the "loss of absolute force": by (3.1) this is

$$|m_1 v_1| - |m_1 u_1| + |m_2 v_2| - |m_2 u_2| = (m_1 + m_2)|v| - (|m_1 u_1| + |m_2 u_2|)$$

$$= |m_1 u_1 + m_2 u_2| - (|m_1 u_1| + |m_2 u_2|). \tag{3.4}$$

---

tion of a remark in Newton's *Principia*. In connection with MacLaurin's Section IV, we note that Mariotte had considered some cases of oblique collisions (see Proposition II in the second part of [74]).

In this corollary u_1 and u_2 have opposite signs and $|m_1 u_1| \geq |m_2 u_2|$, assuming this condition from the first corollary, as MacLaurin apparently intends. The last expression in (3.4) therefore becomes

$$|m_1 u_1| - |m_2 u_2| - (|m_1 u_1| + |m_2 u_2|) = -2|m_2 u_2|,$$

which corresponds to the loss of $2Bu$ stated by MacLaurin.

Proposition III and its corollaries are concerned with the collision of perfectly elastic bodies, for which we have the equations (Introduction, equations (i) and (iv), pp. 49–50)

$$m_1 v_1 + m_2 v_2 = m_1 u_1 + m_2 u_2,$$
$$v_2 - v_1 = u_1 - u_2,$$

with solution

$$v_1 = \frac{m_1 u_1 - m_2 u_1 + 2m_2 u_2}{m_1 + m_2}, \quad v_2 = \frac{m_2 u_2 - m_1 u_2 + 2m_1 u_1}{m_1 + m_2}. \tag{3.5}$$

Hence for the changes of momentum we have

$$m_1 v_1 - m_1 u_1 = \frac{m_1^2 u_1 - m_1 m_2 u_1 + 2m_1 m_2 u_2 - m_1(m_1 + m_2)u_1}{m_1 + m_2}$$
$$= \frac{2m_1 m_2 (u_2 - u_1)}{m_1 + m_2},$$
$$m_2 v_2 - m_2 u_2 = \frac{2m_1 m_2 (u_1 - u_2)}{m_1 + m_2}.$$

As MacLaurin notes in Proposition III, these changes are twice the corresponding changes for the case of perfectly hard bodies, which are given in (3.2) and (3.3).

Corollary III of Proposition III deals with the case where $u_2 = 0$, for which we obtain from (3.5)

$$v_1 = \frac{(m_1 - m_2)u_1}{m_1 + m_2}, \quad v_2 = \frac{2m_1 u_1}{m_1 + m_2}. \tag{3.6}$$

Moreover,

$$m_2 v_2 - m_1 u_1 = \frac{2m_1 m_2 u_1 - m_1(m_1 + m_2)u_1}{m_1 + m_2} = \frac{m_1(m_2 - m_1)u_1}{m_1 + m_2},$$

from which we see that, if $m_2 > m_1$, then $|m_2 v_2| > |m_1 u_1|$, that is to say, the larger mass, which was at rest before the collision, now has momentum with larger magnitude than the smaller mass had before the collision.

MacLaurin next states a result concerning successive collisions of eleven perfectly elastic bodies with masses m_0, $10 m_0$, $10^2 m_0$, ... , $10^{10} m_0$. They are separated and arranged in this order in a row; all are at rest except for the first body, which moves towards the second with velocity u_0. For the first collision we have

$$m_1 = m_0, \quad m_2 = 10m_0, \quad u_1 = u_0,$$

so that by (3.6)

$$v_2 = \frac{2}{11} u_0, \quad v_1 = -\frac{9}{11} u_0.$$

Then in the second collision

$$m_1 = 10m_0, \quad m_2 = 10^2 m_0, \quad u_1 = \frac{2}{11} u_0,$$

which leads to

$$v_2 = \left(\frac{2}{11}\right)^2 u_0, \quad v_1 = -\frac{9}{11} \times \frac{2}{11} u_0,$$

and for the third

$$m_1 = 10^2 m_0, \quad m_2 = 10^3 m_0, \quad u_1 - \left(\frac{2}{11}\right)^2 u_0,$$

giving

$$v_2 = \left(\frac{2}{11}\right)^3 u_0, \quad v_1 = -\frac{9}{11} \left(\frac{2}{11}\right)^2 u_0,$$

and so on. After the final collision the eleventh body will have momentum

$$10^{10} m_0 \left(\frac{2}{11}\right)^{10} u_0 = \left(\frac{20}{11}\right)^{10} m_0 u_0 = (394.796\ldots)m_0 u_0.$$

Note that, as MacLaurin points out, the sum of the momenta after all the collisions have taken place is just the initial value $m_0 u_0$:

$$-m_0 \frac{9}{11} u_0 - 10 m_0 \frac{9}{11} \times \frac{2}{11} u_0 - 10^2 m_0 \frac{9}{11} \left(\frac{2}{11}\right)^2 u_0$$

$$\ldots - 10^9 m_0 \frac{9}{11} \left(\frac{2}{11}\right)^9 u_0 + \left(\frac{20}{11}\right)^{10} m_0 u_0$$

$$= m_0 u_0 \left(\left(\frac{20}{11}\right)^{10} - \frac{9}{11} \left(1 + \frac{20}{11} + \left(\frac{20}{11}\right)^2 + \ldots + \left(\frac{20}{11}\right)^9 \right) \right)$$

$$= m_0 u_0 \left(\left(\frac{20}{11}\right)^{10} - \frac{9}{11} \times \frac{1 - \left(\frac{20}{11}\right)^{10}}{1 - \frac{20}{11}} \right) = m_0 u_0. \tag{3.7}$$

The "very learned author" whose arguments concerning perpetual motion are now attacked by MacLaurin is as before 'sGravesande. Section 1 of his pamphlet *Remarques touchant le mouvement perpétuel* [47] is entitled "Preuves de la possibilité du Mouvement perpétuel, en supposant que la force du corps en mouvement est proportionelle à la masse multipliée par la vitesse." It includes the example of the eleven balls, but also gives no details of the calculations. He asserts that almost 800 degrees of force have been generated by the application of just 1 degree of force to the first ball – this corresponds to the sum of the *moduli* of the terms on the left-hand side of (3.7), namely, $2(\frac{20}{11})^{10} - 1 \approx 788.6$; however, as MacLaurin points out, the directions have to be taken into account. Convinced of the validity of his arguments, 'sGravesande concludes by asserting:

> The only means of answering the arguments which have just been proposed for the possibility of perpetual motion is to deny, along with Mr. Leibnitz, the principle on which they are founded, that the forces of bodies are in the ratio of the products of their masses by their velocities.

Thus he appears to be saying that, if perpetual motion is impossible, then the laws of collisions based on momentum considerations cannot be valid.

In Proposition IV, the final result of this section, MacLaurin states the law of restitution for general e. As an illustration he adds the resulting expression for our v_1 in the case where the two bodies are moving towards each other and $e = \frac{15}{16}$, which corresponds to an experiment of Newton's involving the collision of glass spheres (see p. 23 of [81] ([88]), p. 21 of [84]). Solving equations (i) and (ii) (Introduction, p. 49) produces

$$v_1 = \frac{(m_1 - em_2)u_1 + m_2(1 + e)u_2}{m_1 + m_2},$$

from which we obtain MacLaurin's expression on putting

$$u_1 = V, \quad u_2 = -u, \quad m_1 = A, \quad m_2 = B, \quad e = \frac{15}{16}.$$

Note on Section IV (pp. 66–68). The following geometrical principle is implicit in MacLaurin's solution of the Problem and its Corollaries.

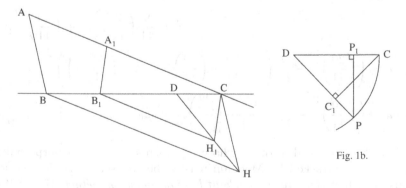

Fig. 1a.

Fig. 1b.

In Fig. 1a we have a parallelogram ABHC and a given point D on BC (cf. Fig. 1, p. 66). From any point H_1 on DH draw H_1B_1 parallel to AC to meet BC in B_1 and from B_1 draw B_1A_1 parallel to CH_1 to meet AC in A_1. Considering similar triangles and opposite sides of parallelograms we obtain

$$\frac{DB_1}{DB} = \frac{B_1H_1}{BH} = \frac{A_1C}{AC}, \quad \text{and so} \quad \frac{DB - DB_1}{DB} = \frac{AC - A_1C}{AC},$$

that is[43]

$$\frac{B_1B}{DB} = \frac{AA_1}{AC}, \quad \text{or equivalently} \quad \frac{BB_1}{AA_1} = \frac{BD}{AC}.$$

MacLaurin chooses D to be such that

$$\frac{\text{velocity of body } A}{\text{velocity of body } B} = \frac{AC}{BD},$$

and so it follows that A_1 and B_1 represent positions of the centres of the bodies, which started at A and B, after the same time t unless the motion has been disturbed by an earlier collision. We can therefore trace out the motion up to collision by letting H_1 vary along DH from H. The length of CH_1 is equal to the distance between the centres at time t, so the bodies will meet when CH_1 is equal to the sum s of their radii. To determine such a point H_1 MacLaurin draws a circular arc with centre C and radius equal to s, obtaining the point L in Fig. 1 (p. 66) for H_1 with A_1 at R and B_1 at N.

Corollary II is clear from the above analysis: if the circular arc does not cut DH, then the distance between the centres is always greater than s and so the bodies cannot meet; if DH is tangential to the arc, the bodies only touch in passing. Referring to Fig. 1b in which P lies on the circle centre D radius DC, we recall that in the standard terminology of the time the sine of the angle CDP with DC as radius is the length PP_1, or equivalently the length CC_1. If P lies on DH, then CC_1 is just the perpendicular distance of C from DH; if this is at least as big as s, then DH either misses the arc or is tangential to it, so that, as MacLaurin states in this corollary, there will be no collision when the sine "is not less than the sum of the semidiameters."

The significance of the second point l in which the arc meets DH (Fig. 1, p. 66) is also clear from the above. Referring to Fig. 1a we see that the minimum value of CH_1 is the perpendicular distance of C from DH, which in Fig. 1 is less than s; the point at which CH_1 is first equal to s is L, after which it decreases to its minimum and then increases, taking on the value s again at l. The corresponding positions are shown in Fig. 1c below, but of course this configuration cannot occur because of the earlier collision – the distance between the centres cannot be less than s.

MacLaurin refers to the point l in Corollary I, a casual reading of which might suggest that Fig. 1c in fact represents the collision when body A starts

---

[43]The operation here is referred to as "by division" in MacLaurin's discussion. It is the ratio operation of *dividendo inverse*: $a : b = c : d \Rightarrow (b - a) : b = (d - c) : d$.

from the point F on AC produced such that AC = CF, its velocity now being in the opposite direction.[44] But this cannot be the case: in the figure, CD is greater than s and, when the centre of body A is at C, the centre of body B is at D; therefore, when the collision takes place, the centre of body B must be to the right of D; however, the construction puts this at the point N to the left of D. I believe that what MacLaurin intended in Corollary I is the following. Consider Fig. 1d below, which is the diagram for the new motion corresponding to Fig. 1a, except that the lengths have been halved. The parallelogram ABHC is now replaced by the parallelogram FBHC.

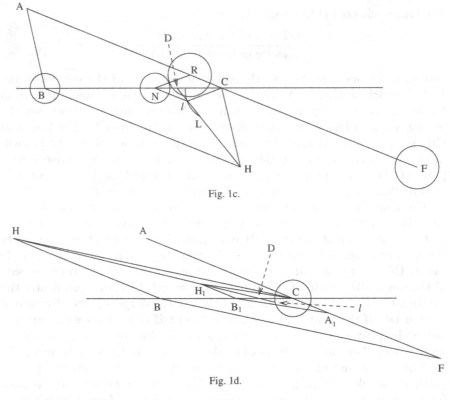

Fig. 1c.

Fig. 1d.

In this case the point at which CH_1 first becomes equal to s will be the *upper* intersection (l) of DH with the arc, whereas in the previous situation it was the *lower* intersection (L).

MacLaurin observes in Proposition V that if velocities are resolved into components parallel and perpendicular to NR (Fig. 1, p. 66) it is only the former which are affected by the collision; moreover, since by our discussion above

---

[44]In MacLaurin's published Fig. 1 the point F is located erroneously on BC produced.

$$\frac{AR}{BN} = \frac{\text{velocity of body } A}{\text{velocity of body } B},$$

and $\overrightarrow{AR} = \overrightarrow{AQ} + \overrightarrow{QR}$ and $\overrightarrow{BN} = \overrightarrow{BM} + \overrightarrow{MN}$, we deduce that the collision takes place with initial velocities such that

$$\frac{\text{component of velocity of body } A \text{ parallel to NR}}{\text{component of velocity of body } B \text{ parallel to NR}} = \frac{QR}{MN}, \qquad (4.1)$$

$$\frac{\text{component of velocity of body } A \text{ perpendicular to NR}}{\text{component of velocity of body } B \text{ perpendicular to NR}} = \frac{AQ}{BM}. \qquad (4.2)$$

The components of velocity parallel to NR after the collision are to be calculated using the results of Section III as if the bodies had been moving along NR with the components described in (4.1) as initial velocities: the new components are represented by Rg and Nm in MacLaurin's terminology. Finally, MacLaurin assumes that all these components are known in units which make the initial velocity of the body A equal to $\overrightarrow{AR}$. Then in MacLaurin's notation (see Fig. 2, p. 68), since the components in (4.2) do not change,

$$\text{velocity of body } A \text{ after collision} = \overrightarrow{AQ} + \overrightarrow{Rg} = \overrightarrow{Rq} + \overrightarrow{qa} = \overrightarrow{Ra},$$

$$\text{velocity of body } B \text{ after collision} = \overrightarrow{BM} + \overrightarrow{Nm} = \overrightarrow{Nl} + \overrightarrow{lb} = \overrightarrow{Nb}.$$

In this way the direction of motion after the collision is determined for both bodies.

Note that the L and l in Fig. 2 are not the same as the points so labelled in Fig. 1. Other aspects of Fig. 2 are also unsatisfactory. The angle between the lines AR and BN is significantly different from the corresponding angle in Fig. 1, as a result of which $\overrightarrow{QR}$ has opposite directions in the two figures. Also the radii of the two bodies have been made the same in Fig. 2 (cf. Appendix II.2, p. 80), which does not seem to be required in the discussion.

There is some confusion in the second paragraph of Proposition V when MacLaurin goes from *velocities* represented by AR and BN to the resolution of *forces* represented by these lines. Does he now mean momentum rather than velocity, in which case he would require the bodies to be the same, as suggested in Fig. 2, if the same lines could also represent momentum? We note finally that 'sGravesande deals with oblique collisions in [48]; some of his diagrams have a marked similarity to MacLaurin's Fig. 2.

Appendix II

II.1. Notice prefixed to the published version of MacLaurin's essay (translation)

NOTICE.

The Academy believes that it must draw attention to the fact that people have not taken sufficient care to confine themselves within the bounds of the question which it had proposed: there were even some authors who did not discuss it, and who substituted another one for it. The Laws of the Collision of perfectly hard Bodies were asked for, without consideration of whether these bodies exist. However, it is only the Laws of the Collision of elastic Bodies which have been given in some of the submitted Memoirs; amongst these there are some of excellent quality, and above all one which has as a motto, In magnis voluisse sat est,[45] in which the author demonstrates much skill in Geometry and much acuteness in the resolution of the most difficult problems.

Since the Laws of the Collision of Bodies and of the transmission of motions are not the same in elastic bodies as in bodies which are infinitely hard, or inflexible, the estimation of the forces, which is at the present time a much debated question, and where perhaps there has been some misconception up till now, can also not be the same in the two cases. One author can have done this estimation properly in the first case, while another can have given a different but valid version of it in the second.

The work which has won the Prize is that of Mr MacLaurin, Professor of Mathematics in the University of Aberdeen and Fellow of the Royal Society of London.[46]

---

[45]This was the essay [13] by Jean Bernoulli to which reference was made above. The quotation comes from Propertius, *Elegies* II.x.5 and may be translated as, "in great endeavours even to have had the will is enough" ([89], p. 530:1).

[46]Corrections from the *Errata* are incorporated here; in particular, the author's name appears as Maclorrins in the text.

II.2. MacLaurin's diagrams from [1] (courtesy of Glasgow University Library, reproduced with permission)

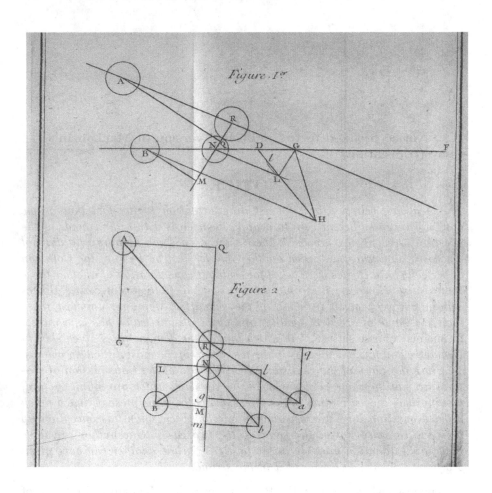

II.3. Extract from 'sGravesande's paper [44]: material between the statement of Proposition IX and its proof (translation)

In order to show first of all that it requires less effort to give a certain degree of velocity to a body, than to increase by the same degree the velocity of an equal body but in motion, it suffices to remark that it would require the same effort in the two cases, if in the second the moving cause were transported with the velocity which the body had before the increase, which does not happen without effort. Let us imagine two men A and B each holding a ball; we suppose that the two balls are equal; A is at rest; B is on a boat with which he is transported: this gives the velocity which the boat has to the ball which B is holding. The two men throw their balls by making equal efforts; then the increase in the velocity of the ball which B has thrown is equal to the whole velocity of the ball which A has thrown. In order to give to this last ball its velocity, the effort which A has made is enough, but to increase the velocity of the other ball, over and above an equal effort by B, it is necessary that B be transported.

II.4. Response by 'sGravesande to MacLaurin's example in the second paragraph of Article I of Section III (translation)

(See [46] or [49], Vol. I, Part I, pp. 262–265 for the original.)

This reasoning appears to be founded on very simple principles and would be conclusive if these principles were indeed such; but it is necessary to examine if attention has been paid to everything which must be considered.

It is supposed as clear that the man who is on the ship communicates 8 degrees of velocity to the body A by making precisely the same effort as the man on the shore, who gives the same velocity to the body B. This, however, is not exactly true; if we put a man on a plank or in a small, light launch, and if he throws some heavy mass, for example something of one hundred or two hundred livres, it will be seen whether, with a specific effort, he will be able to communicate to it as much velocity, as if he were located on a fixed base.

It is true that the heavier the ship, the less the difference will be between the efforts which give the velocity 8 to each of the bodies A and B; however there will always be some: and this difference, however small we make it by increasing the ship, will always be sufficiently large to remove from the objection its entire force and to confirm the opinion which I am defending.

This is what we will try to prove after having presented some principles.

I. A spring which relaxes makes all its effort in one direction, provided it cannot go back.

Proof from experiment. A compressed spring, which, when placed between two equal bodies, communicates to each of them a certain degree of velocity, throwing one to the right and the other to the left, if, being fixed to an immovable obstacle, it pushes the two bodies together in the same direction, will communicate to each of them the same degree of velocity as in the first case.

II. A spring which is transported while it relaxes communicates all its effort in the direction towards which it is transported.

This is clear, since a transported spring is a spring which cannot go back.

III. The force, that is to say, the capacity to act, which a body acquires, is equal to the action which communicates it.

The cause is proportional to the effect.

IV. In order to determine the total action which serves to give force to a body, it is not sufficient just to consider the immediate action of the moving cause on the body, but it is necessary to add to it that which serves to transport the moving cause, if this last action has no other effect than to put the moving cause in the state of acting with greater effectiveness.

Proof from experiment. The effect of two springs is the same, whether they act together, one beside the other, or one pushes the other while the latter relaxes.

V. A spring, at rest between two bodies, on relaxing communicates to them velocities which are in the inverse ratio of the masses.

This is consistent with experiment and, besides, it is not contested.

Now that these principles have been set down, we suppose that the body which is on the ship, instead of being thrown by a man, is pushed by a spring attached to the ship in such a way that it cannot go back without causing the whole ship to go back. This changes nothing in the reasoning which we are examining, the action of the spring being analogous to that of the man; but it is more regular and it serves to make the calculation more meaningful.

The body A applied to the taut spring has along with the ship 2 degrees of velocity; it has 4 degrees of force. The spring relaxes and communicates to the body 8 degrees of velocity in the ship, that is 64 degrees of force in the ship. Up to here we are in agreement. In the air the body has 10 degrees of velocity; it therefore has 100 degrees of force: this is what we again agree. But, here is the difficulty; it is said that this is not possible, because the body, which had 4 degrees of force, has only received 64. Now this is what I do not see. Indeed the body has only received 64 degrees of force to act upon an obstacle transported with 2 degrees of velocity; but, to consider everything, the body has acquired 96 degrees of force, as we are going to try to prove.

In order to perform the calculation, it is necessary to know the mass of the ship; but, whatever we put as this mass, the result of the calculation is the same.

Let us suppose this mass to be a thousand times greater than that of the body A: we can take the number as we want.

We have a spring between two bodies, between the ship to which it is attached and between the body A; this spring is at rest with respect to these two bodies. Consequently, on relaxing, it communicates to them velocities which are in the inverse ratio of the masses (V). It communicates 8 degrees of velocity to A, which is why it communicates to the ship $\frac{8}{1000}$ or $\frac{1}{125}$ degrees of it.

The effort of the spring must be measured by considering that it was at rest between the two bodies if they had been at rest. Therefore I multiply 64, the square of the velocity communicated to A, by its mass 1 and I have 64. I multiply $\frac{64}{1000000}$, the square of the velocity communicated to the ship, by its mass 1000 and I have $\frac{64}{1000}$ or $\frac{8}{125}$, and the total action of the spring is equal to $64\frac{8}{125}$; for this spring between the two bodies at rest would have produced such a force.

We are concerned here with a transported spring, which makes all its effort in the direction towards which it is pushed (II); this is why it communicates to the body A a force which is equal to $64\frac{8}{125}$.

The spring of which we are speaking is a transported moving cause, and the action of the ship, which pushes the spring while it is acting, makes it more capable of acting on the body, which, without that, would avoid part of the action of the spring as a result of its own movement. For this reason the action of the ship, by which the moving cause is transported, also communicates itself to the body A (III, IV). It is therefore necessary to determine this action, which is equal to the force which the ship has lost as a result of this action, since the effect is proportional to its cause.

This lost force is found if the force after the action is subtracted from the force before the action; what is left is the lost force and this added to $64\frac{8}{125}$, which we have already, will give everything which the body gains as a result of the actions of the ship and the spring combined together (IV).

The mass of the ship, 1000, multiplied by 4, the square of its velocity before the spring relaxes, gives 4000 degrees of force before the action.

The velocity $\frac{1}{125}$, which the spring has communicated to the ship, has direction opposite to that of the initial velocity, which consequently has been diminished, and $1\frac{124}{125}$ is left, whose square $3\frac{15126}{15625}$ multiplied by the mass 1000 gives the force after the action, $3968\frac{1000}{15625}$ or $3968\frac{8}{125}$; when this has been subtracted from 4000, it gives $31\frac{117}{125}$ for the effort of the ship on the spring, to which must be added $64\frac{8}{125}$, which is the effort of the spring, in order to have the whole force which the body A has gained; if we add to this 4, the force of the same body before the action of the spring, we will have the total force; now the sum of these numbers is 100 and not 68.

If one wishes to take the trouble to carry out the calculation algebraically, it will be seen that the demonstration is general and that the force acquired is always proportional to the action which communicates it; this will not be true if we use a different method of measuring the force, unless we call into doubt a matter which is nevertheless consistent with experiment, as we have seen above (IV); it is that the agent which transports and pushes the moving cause during its action acts with it on the same body.

II.5. Proposition XIV from 'sGravesande's paper [44] along with its proof (translation)

The force lost in the collision of two nonelastic bodies is the same whatever the absolute velocities of these two bodies may be, provided their relative velocity is the same.

The motion of the two bodies is composed of their common motion and their relative motion. It is clear that the first, in whatever manner it may be changed, cannot change the action of one body on the other; thus this action is always the same as long as the relative velocity does not change. It is on this action or effort of the bodies, the one against the other, that the flattening or breaking down of the parts depends, which consequently will be the same if the relative velocity is the same. This is consistent with known experiments.

In the collision there is no force lost except that which is required to flatten or break down the parts. Consequently, this lost force is the same when the flattening or breaking down of the parts is the same, that is to say, in all the cases in which the relative velocity of the two bodies is the same.

Part III

MacLaurin on the Tides:
De Causa Physica Fluxus et Refluxus Maris
(Royal Academy of Sciences, Paris, 1740)

Part III Contents

Introduction to Part III

The prize topic proposed by the Royal Academy of Sciences for 1740 was the tides: *le Flux et Reflux de la Mer*. The winners, in the order in which their essays were published by the Academy, were Father Antoine Cavalleri [23], Daniel Bernoulli [12], Colin MacLaurin [68] and Leonhard Euler [38]. Cavelleri's work was based on the Cartesian idea of vortices;[47] the other three were founded on Newtonian principles and were subsequently reproduced in the "Jesuit Edition" of the *Principia* (1739–1742) [86] as illustrations of the Newtonian philosophy. Newton had already discussed in its earlier editions the tidal forces on the Earth which resulted from the gravitational attraction of the Sun and of the Moon; his work was of course the starting point for MacLaurin's investigations.

Intimately connected with the study of the tides is the problem of the figure of the Earth, which was also discussed by Newton. For MacLaurin the approach to both problems was the same, namely, the equilibrium of a spheroidal[48] fluid mass held together by the mutual attraction of its particles according to the inverse square law of attraction and acted upon by certain external forces: for the tides these are the attractions of the Sun and of the Moon; for the figure of the Earth there is just one external force, the centrifugal force resulting from the Earth's rotation about its axis. In the case of the tides the fluid sphere is elongated along an axis by the action of the external forces, so we have an *oblong* spheroid, while for the figure of the Earth the centrifugal force causes a flattening at the poles, producing an *oblate* spheroid; the resulting mathematical formulae are significantly different in the two cases.

Throughout the 1730s and early 1740s the study of the figure of the Earth was a major topic which occupied many authors, including James

---

[47] According to [50] the judges had agreed that one of the prizes should be reserved for an essay based on Cartesian principles. Cavalleri, a Jesuit professor of mathematics and theology, had been awarded prizes in 1738 and 1739 by the Bordeaux Academy for essays on other physical topics. Some discussion of his essay on the tides will be found in §4 of [5].

[48] For MacLaurin a *spheroid* is the solid of revolution obtained by rotating an ellipse about one of its axes. The spheroid is *oblong* (sphaerois oblonga) or *oblate* (sphaerois oblata) according as the rotation is about the major (transverse) or the minor (conjugate) axis.

Stirling [97], Alexis-Claude Clairaut [25–27], Colin MacLaurin [68,69] and Thomas Simpson [96]; their contributions are discussed by Todhunter [103] and recently by Greenberg [50] (see also [105], Chapter 5 and Appendix B). The theoretical work was supported by two expeditions organised by the Royal Academy of Sciences to determine by surveying the lengths of a degree of latitude in the polar region (Lapland, 1736–1737) and at the equator (Ecuador (at that time part of Peru), 1735–1744) (see [103], Chapters VII and XII); the very successful northern expedition was led by Maupertuis [75] and included Clairaut, who communicated his first paper [25] on the figure of the Earth to the Royal Society from the expedition's base at Torneo.

MacLaurin was certainly working on the figure of the Earth by 1738; this is known from the Stirling–MacLaurin correspondence (see, for example, three letters of May 1738 ([111], pp. 80–88, [77], pp. 293–303)). Whether he had applied his theory to the study of the tides prior to the announcement of the prize topic does not seem to be known, although we can be sure that he was already completely familiar with Newton's contributions. Stirling, MacLaurin and Simpson, and perhaps also Clairaut, each have some claim to priority in the resolution of the problem of the figure of the Earth. Simpson asserts in the Preface to [96]:

> I must own that, since my first drawing up this Paper, the World has been obliged with something very curious on this Head, by that celebrated Mathematician Mr. Mac-Laurin, in which many of the same Things, are demonstrated. But what I here offer was read before the Royal Society *, and the greater Part of this Work printed off, many Months before the Publication of that Gentleman's Book; for which Reason I shall think myself secure from any Imputations of Plagiarism, especially as there is not the least likeness between our two Methods.
> ......
>
> * It was read before the Royal Society in *March* or *April*, 1741, and had been printed in the Philosophical Transactions, had I not desired the contrary.

Certainly, Simpson's methods, based on clever manipulations of infinite series, are quite different from MacLaurin's geometrical approach. However, as noted by Todhunter [103, Article 247], the gist of MacLaurin's work, in particular his Fundamental Theorem on the equilibrium of fluid bodies, had already appeared in his prize essay, predating the *Treatise of Fluxions* [69], to which Simpson alludes. The editors of [86], Father Thomas Le Seur and Father François Jacquier, gave MacLaurin special mention in their introduction to the three winning entries reproduced there, pointing out that he had in fact solved in his 1740 essay a problem which they had proposed in their Notes on Newton's Proposition XIX of Book III but had been unable to resolve: this was the spheroidal shape of the Earth. Stirling's claims were promoted at the Royal Society, not by himself, but by John Machin; however, they are

based on unpublished material, which is now probably lost.[49] Clairaut wrote
the period's definitive text on the figure of the Earth [27] (1743); in place
of the methods of his earlier papers he adopted in part those of MacLaurin,
which had greatly impressed him.[50] Several letters from Clairaut to MacLaurin are extant (see [77]); from one of these (Letter 174 of [77]) we learn that
Clairaut had been one of the judges for the 1740 competition. In [69] MacLaurin went on to consider nonhomogeneous layered spheroids, which were also
investigated by Clairaut; this aspect of MacLaurin's work and its influence
on Clairaut are discussed in [50] (see especially §8.2). Patrick Murdoch, who
was MacLaurin's friend and the editor of his posthumous *Account of Sir
Isaac Newton's Philosophical Discoveries* [70] ([73]), published [80], in which
he established by his own methods some of MacLaurin's results. Many great
names make an appearance in the further study of the figure of the Earth, including Lagrange, Legendre and Laplace, but let us note finally the continued
Scottish input represented by the highly regarded, if sometimes controversial,
19th-century work of James Ivory (1765–1842) (see [33, 34, 103]).

Newton's law of gravitation effectively asserts that the force of attraction
between two particles of masses m_1 and m_2 and at distance r apart is given
by

$$\frac{Gm_1m_2}{r^2}, \tag{1}$$

where G is a constant; the force of attraction on one particle acts in the
direction from that particle to the other. This law governs aspects of the
motion of the Moon about the Earth and of the Earth about the Sun. It also
determines, in part, the tidal forces on the Earth caused by the Sun and the
Moon. Although the gravitational attraction between the Sun and the Earth
is much greater than that between the Moon and the Earth, it turns out
nevertheless that the Moon has a greater influence than the Sun on the tides
of the Earth's waters. This is because it is the *variation* of the attraction over
the Earth's surface that determines the tides rather than its magnitude; the
variation, measured by differences, depends on

$$\left| \frac{d}{dr}\left(\frac{Gm_1m_2}{r^2} \right) \right| = \frac{2Gm_1m_2}{r^3}. \tag{2}$$

The relevant quantities for the Earth, Moon and Sun are such that (1) is
greater for Earth–Sun than for Earth–Moon, while the reverse is true in the
case of (2) (for the calculations see, for example, [55], pp. 108–110). Newton
and MacLaurin both use the fact contained in (2) that, at the great distances

---

[49]See Chapter 5 of [105] for details and some relevant material extracted from
Stirling's notebooks.

[50]Clairaut wrote: "... *j'ai jugé à propos de traiter en particulier de la figure des
sphéroïdes homogènes, et d'abandonner ma méthode, quant à ces sphéroïdes, pour
suivre celle que M. Mac Laurin vient de donner dans son excellent Traité des
fluxions. Cette méthode m'a paru si belle et si savante, que j'ai cru faire plaisir
à mes lecteurs de la mettre ici.*"

involved, tidal forces are approximately inversely proportional to the cubes of the distances.

Now let us turn to a brief description of the contents of MacLaurin's 1740 essay on the tides – detailed analyses of individual results and items will be found in the Notes[51] placed at the end of the translation. Section I describes some of the observed phenomena associated with the tides and Section II is largely an outline of how the Moon, the Sun and the motion of the Earth appear to influence the tides. Also in Section II MacLaurin states certain approximations which are implicit in the *Principia* and which form the basis of some of his later applications; here the inverse square law (1) is used: the lengths of the lines chosen to represent certain forces are taken in inverse proportion to the squares of the distances involved (see Fig. 1 and the associated discussion, pp. 102–103).

A number of technical terms arise in these two sections, some of which are important later on. It is convenient to note the important ones here:[52] the *syzygies* occur when the Earth, Moon and Sun are aligned;[53] in the *quadratures* the line from the Earth to the Moon is perpendicular to the line from the Earth to the Sun; the Moon's *declination* is its angular height relative to the Earth's equatorial plane (measurement as for latitude). Concerning the heights of tides we should note that these are generally given in the old French units: 12 *lignes* = 1 *pouce*, 12 *pouces* = 1 *pied* ≈ 1.066 feet; the terms *pied* and *pouce* are commonly written in English as *Parisian* (or *Paris*) *foot* and *inch*.

Section III contains the main results of MacLaurin's essay. Proposition I, which MacLaurin also labels *Theorema Fundamentale*, is his famous theorem on the equilibrium of a spheroidal mass of fluid. The proof which MacLaurin presents glosses over some important points; nevertheless, MacLaurin does go a considerable way towards justifying his result, which represented a substantial new insight into the problem. In the Corollaries he applies his Fundamental Theorem to obtain approximate expressions for the ratio of the "difference of the height of the water" at the poles and at the equator to the Earth's mean diameter in the syzygies and the quadratures; this difference is of course the difference in the lengths of the axes of the generating ellipse for the spheroidal shape taken on by the fluid Earth under the combined effects of the Moon and the Sun. Proposition I is preceded by four Lemmas concerned with geometrical and gravitational matters which are required in the development. Of these, Lemma IV is the most significant: here we find MacLaurin ingeniously exploiting the geometry and symmetry of the spheroid to determine components of attraction; in following through MacLaurin's ar-

---

[51]References to notes are usually given in abbreviated form: for example, NLI indicates the Note on Lemma I and NPIII denotes the Note on Proposition III.

[52]Distances between bodies are understood to go from centre to centre.

[53]In Corollary 4 of Proposition I MacLaurin distinguishes two separate cases of syzygies: where the Moon is in *conjunction* with the Sun (they are on the same side of the Earth), and where they are *opposed* or in *opposition* (on opposite sides).

gument I have the distinct impression that he is essentially evaluating the triple integrals by which these components are represented nowadays (see Appendix III.3 (pp. 201–202)). This theme continues in Lemma V, which expresses by means of a certain integral the attraction at the poles for a general class of figures and compares this with the attraction of a related sphere. In Proposition II MacLaurin applies this nicely to determine the attraction at the poles of an oblong spheroid. Lemma VI and Proposition III deal in a similar way with the attraction at the equator of an oblong spheroid, for whose determination, as MacLaurin notes, the method is less obvious. The results of Propositions II and III are combined in Proposition IV along with MacLaurin's earlier considerations of the ratio of the difference in the lengths of the fluid Earth's axes to its mean diameter to give an expression for this ratio in terms of the tidal forces caused by the Sun and the Moon and the mean gravity over the Earth's surface; the effect of the Earth's rotation is ignored. The section ends with a discussion in Proposition V of the tidal force due to the Sun alone; here MacLaurin quotes various results and data from Newton. MacLaurin's Fundamental Theorem applies to both oblong and oblate spheroids but, as noted above, it is the oblong spheroid which is relevant in the applications to the tides; however, in a series of Scholia to his Propositions, MacLaurin states corresponding results for oblate spheroids, so that, implicit in his essay, is an outline of aspects of his theory of the figure of the Earth.[54]

Section IV begins with a discussion of the effect of the Earth's rotation on the tides and a calculation to determine where the height of the water is least in the syzygies when the rotation is taken into account. In Proposition VI MacLaurin applies his formula from Proposition IV to consider the tidal forces on the Moon due to the effects of the Sun and to compare the ellipse corresponding to the resulting shape of a fluid Moon with the observed ellipse of the lunar orbit. Proposition VII points out that, since the velocity of a place on the surface of the rotating Earth decreases as we move north or south from the equator, the directions of tides will be affected as the moving water changes latitude. MacLaurin now allows the Earth to have a more general shape (Fig. 11, p. 131): all sections through the axis Aa have to be elliptical, but the equatorial section $ABab$ is now an ellipse and not necessarily a circle. In Proposition VIII MacLaurin states a few terms of a series giving the attraction at A from this solid and indicates how by use of this series and a related series for the attraction at B the rise of the water can be determined. However, he dismisses this approach as being of little use. The section concludes in Proposition IX with some observations, partly from Newton, concerning the tidal forces produced by the Moon and the possibility of only one tide in a day at certain places.

---

[54]For MacLaurin's proofs see [69], especially Articles 641 and 646. Some details are also given in my Notes.

At the end of the essay there is an appendix of three *Remarks* (Annotanda) from MacLaurin. In the first of these he notes an error which he had made in the expression given in Proposition IV for the ratio of the rise of water to the mean diameter and gives an improved version. The second Remark gives an indication of how he derived the series stated in Proposition VIII. Finally he notes a variant of a procedure given by Newton for comparing the tidal forces produced by the Moon with those produced by the Sun. I believe that the *Remarks* did not form part of the original submission but were sent in afterwards to be published as an appendix to it: there is a single diagram (for Remark II) which is of poor quality in comparison with the carefully engraved diagrams (Figs 1–11) elsewhere in the essay, suggesting that this material may have arrived after the production process was underway.[55]

A number of typographical and other errors appear in the published version [68] of MacLaurin's essay; I have corrected without comment the obvious typographical errors and in the case of the small number of more serious errors, I have discussed these in my Notes. All seem to be easily remedied and none has a detrimental effect on the general merit of the ideas put forward in the essay. Patrick Murdoch tells us in *An Account of the Life and Writings of the Author*, which is contained in [70] ([73]), that MacLaurin "happened to have only ten days time to draw up this paper, and could not find time to transcribe a fair copy, so that the *Paris* edition of it is incorrect; but he afterwards revised the whole and inserted it in his Treatise of Fluxions."[56] We should also note that, while they claim to have corrected the errors appearing in the *Paris* edition, the editors of [86] have in fact allowed several new errors to creep in to their version of MacLaurin's text.

To some extent the essay and the discussion of the figure of the Earth and the tides in [69] are complementary: the emphasis in [69] is on the figure of the Earth and the applications to the tides are given in rather contracted form; some items, however, are taken rather further than in the essay (see, for example, Articles 690–691). The relevant material in [69] is contained in its Chapter XIV, which begins with several Articles under the heading *Of the Ellipse considered as the Section of a Cylinder*. Here MacLaurin derives properties of the ellipse, some of which are required in the study of the figure of the Earth and the tides, from corresponding properties of the circle by means of a projection method; he also considers for his applications the sections of spheroids by planes (Article 633). I have outlined the projection method in Appendix III.1 (pp. 197–199) and I have indicated how it may be applied at appropriate places in my Notes. Appendix III.2 (pp. 199–201) contains a discussion of plane sections of spheroids using coordinates.

Of course there are points of contact as well as major differences among the essays of MacLaurin, Euler and Bernoulli. The distinguishing features of MacLaurin's work would appear to be his Proposition I, or Fundamental

---

[55]Figs 1–11 from [2] are reproduced in Appendix III.6 (pp. 209–210).
[56]MacLaurin discusses the theory of the tides in Articles 686–696 of [69].

Theorem, on the equilibrium of a spheroidal mass of fluid and his geometrical methods. Bernoulli's essay is discussed extensively in [64] by Lubbock, who makes many references to MacLaurin's work and deduces some of the results of MacLaurin, Euler, Stirling and Clairaut in the second part of [64] (see also [4, 50]). As a potentially useful application of his theory, Bernoulli produced tables from which the heights of tides could be predicted when the Moon is at its perigee, mean distance from the Earth or its apogee; the heights are given as functions of the local spring and neap tides for angular separations of the luminaries from 0° to 180° in steps of 10°. Concerning Euler's essay, see the introduction to it by E. J. Aiton in [40] (also [4]) and Part II of the Editor's Introduction by C. Truesdell in [39], which also makes reference to MacLaurin's work. Aiton has in fact written extensively on the tides (see [3–5]): particularly relevant here is [4], which discusses the contributions of Newton, as well as those of Bernoulli and Euler. According to Aiton [4], Euler's main contribution to the study of the tides was his identification of the horizontal component of the disturbing force as the main cause. The Cartesian theory, including Cavalleri's essay, is outlined in [5].

Translation of MacLaurin's Essay

Concerning the Physical Cause
of the Flow and Ebb of the Sea.

by the most learned
Mr MacLaurin
Professor of Mathematics,
from the
Society of the Academy of Edinburgh.

Opinionum commenta delet dies, naturae judicia confirmat.[57]

(p.137) ## SECTION I.

Phenomena.

In times past the philosophers recognised a triple motion of the sea, * daily, monthly and annual; in the daily motion the sea rises up and flows away twice each day, in the monthly motion the tides are increased during the syzygies of the luminaries and are diminished at the quadratures, and finally in the annual motion the tides become greater in winter than in summer: but these phenomena require to be set down a little more accurately.

I. The daily motion of the sea is completed in about 24 solar hours and 48 first minutes,[58] namely the interval of time in which the Moon, having left from the meridian of some place, its motion being observed, comes back to the

---

[57]The quotation is from Cicero's *De Natura Deorum* and has been translated as, "The years obliterate the inventions of the imagination, but confirm the judgements of nature" ([92], pp. 126–127).

* Pliny Book 2, Chapter 99.

[58]This means minutes in the usual sense, the first division of the hour into 60 parts; likewise seconds, or second minutes, correspond to the second division into 60 parts.

same place. Hence the greatest height of the sea is connected with the Moon moving to a given position with respect to the meridian of a given place; but the solar hour in which the tide occurs is delayed from day to day, by almost the same interval by which the Moon's arrival at the meridian of the place is delayed. And this motion is so accurately adapted to the motion of the Moon, that, in accordance with the observations reported by the celebrated Mr Cassini, a calculation may be made of the hour in which the true conjunction or opposition of the Sun takes place, and an equation deduced from the motion of the Moon may be applied in order that the time when the sea will rise up to its greatest height on the day of the New Moon or of the Full Moon may be defined more accurately. Moreover, in estuaries various tides arise seasonally, as Pliny says, which are not at variance with the scheme. Two tides which are produced on individual days, are not always equal; for the early morning tides are greater than the evening tides in winter time, but smaller in summer time, especially at the syzygies of the luminaries.[a]

II. Concerning the monthly motion of the sea there are three things to be noted in particular. (1) The tides become greatest in individual months a little after the syzygies of the Sun and the Moon, they decrease in the transit of the Moon to the quadratures, and they are least a little after. The differences are such that the greatest rise of the entire water to its least rise in the same month is, according to certain observations, as 9 to 5, and in some cases the difference is observed to be somewhat greater. (2) The tides are greater, other things being equal, the smaller the distance of the Moon from the Earth, and that in a greater ratio than the inverse square of the distances, as is inferred from various observations. For example, in the year 1713 the rise of the water in the port of Brest,[b] according to the same distinguished man, was 22 feet 5 inches on 26 February and 18 feet 2 inches on 13 March. The declination of the Moon was almost the same in both cases; in the former the distance of the Moon was 953 parts, in the latter 1032 parts, the mean distance of which is 1000. Now the square of the number 1032 to the square of the number 953 is as 22 feet 5 inches to 19 feet $1\frac{2}{3}$ inches; but the rise of the water in the latter case was only 18 feet with 2 inches. (3) Other things being equal, the tides are greater when the Moon stays in the equatorial circle, and are reduced as the declination of the Moon from this circle grows.

III. The tides become greater, other things being equal, the smaller the distance of the Sun from the Earth; and so they are greater in winter than in summer, other things being equal. However, the difference is far less than that which arises from the varying distances of the Moon. For example, the perigeal distances of the Moon were equal on 19 June 1711 and on 28 December 1712; the rise of the water on the former date was 18 feet 4 inches, and

[a] *Mém. de l'Acad. Royale,* 1710, 1712 & 1713.
[b] *Ibid.*

on the latter 19 feet 2 inches; but the declination of the Moon was a little less in the latter than in the former observation.[a]

Moreover, the tides are different in different places, on account of the different latitude of places, and their location with respect to the ocean from which they are generated, because of the size of the ocean itself, and the nature of the shores and straits, and for other various reasons.

(p.137)

SECTION II.

Principles.

Now that the more distinguished phenomena of the tides of the sea have been briefly reviewed, we proceed to the principles, from which their calculation is to be made. But let it be stated beforehand that this part of philosophy, which investigates and explains the causes of phenomena, is indeed the most noble but at the same time the most difficult. It is the subtlety of nature that it is not to be wondered at if the primary causes escape the ingenuity of philosophers for the most part. They have taken it upon themselves to reveal the laws of all phenomena, and to show us the whole series of causes; they have certainly failed in their great undertakings up to this time. Indeed, very distinguished men have proposed to themselves the development of the most complete philosophy, which, however, is such that it is right to doubt that it is in agreement with human destiny. It is therefore preferable to follow cautiously and slowly the traces of nature itself, having been thoroughly informed by the less fruitful success of so many men. For if we can reduce phenomena to certain general principles, and subject their properties to calculation, we will grasp as a result of these steps some part of the true philosophy; this will indeed be defective or incomplete, if the causes of the principles themselves are not revealed: however, so great is the beauty in the nature of things that that part is far superior to the finest inventions of very acute men.

It is obvious to anyone, or rather to anyone who considers the matter a little, that the motions of the sea are associated with and are similar to the motions of the luminaries, especially those of the Moon. The period of the daily motion of the sea is the same as that of the Moon at the meridian of the place, and the period of the monthly motion is the same as that of the Moon in relation to the Sun; the force of both luminaries in generating motion of the sea in this way is apparent, because the smaller the distances of both are from the Earth, the greater the tides are; and so there are no grounds for doubting that the motion of the sea is connected by some relationship to the motion of the Moon and the Sun. Moreover, we will affirm such things to be forces propagated by the Moon and the Sun (or dependent on these in some way) which raise and depress the water from day to day; they act

---

[a] *Mém. de l'Acad. Royale*, 1710, 1712 & 1713.

together in the syzygies of the luminaries and oppose in the quadratures; they are increased in the smaller distances of both and decreased in the larger distances; they are stronger in a smaller declination of the Moon and weaker in a large declination; and they sometimes cause a greater motion when the Sun and the Moon are sunk below the horizon, than when both are dominant in the upper meridian. There have been very distinguished men who believe that the tide of the sea arises from a certain pressure of the Moon. But they do not make known the cause and the measurement of this pressure, nor do they explain sufficiently clearly how the various motions of the sea can arise from this, much less have they taught (this principle having been laid down) how to reduce those motions to calculation.

The most sagacious Kepler pointed out long ago that the sea gravitates towards the Moon and hence the tide of the sea is set in motion. After he had discovered the laws of gravitation, Newton found that the equilibrium of the sea was disturbed less by its gravitation towards the Moon, than by the unevenness of the force under which the particles of the sea tend to the Moon and the Sun according to their different distances from the centres of these, and he was the first to show how to reduce the motion of the sea to certain laws and calculation. Certainly it has to be admitted that the cause of gravity is not known, or at the very least it is obscure; however, bodies are not less heavy on that account. There may be those who assert that bodies try to reach each other not because of an impulse or an external force, but as a result of a certain innate force; but it is not right to associate their whims with the truth. Others steadfastly take refuge in the immediate power of the supreme Author, but neither is the excessive haste of the latter to be approved, nor the disdain of the former who do not take note of so many testimonies of nature that the cause of gravity is obscure. The force of gravity is so familiar to us, and its measurement held to be well-established, that we almost always use this for the calculation of other forces; the distinguished man has shown with such great clarity that this has dominion in the Heavens, no less than on the Earth, and that it increases and decreases according to a certain law, so that in vain may you seek a greater authority in this lofty and difficult part of philosophy, which is concerned with the causes of things.

Newton showed by an exceptional argument that the Moon is attracted towards the centre of the Earth by a force which (the ratio of the distances having been obtained) agrees entirely with the gravity of terrestrial bodies; thus it may be asserted with equal justification that the Earth is attracted towards the Moon in equal measure. When one body is driven towards another, it certainly does not follow from this that the latter is pushed at the same time towards the former. But what is to be noticed about the gravity of celestial bodies is very well discerned from those things which have been found out about the gravity of terrestrial bodies (and other similar forces); for we are led from the latter to an understanding of the former and the phenomena are entirely similar. A mountain gravitates into the Earth, and if the Earth did not push the mountain with an equal and opposite force, the Earth, pushed by the mountain, would go off to infinity with an accelerated motion. On the other hand the status of any system of bodies (i.e., the motion of its centre of gravity) is necessarily disturbed by every action to which there is no equal and opposite reaction, so that scarcely anything can be said to be lasting or constant in the system if this law does not hold. And since the parts of the Earth always act on each other mutually in such a way that the motion of the centre of gravity of the Earth is not at all disturbed by mutual collisions of bodies or any other forces, located either within or outside the surface, and since the same law holds in magnetic, electric and other forces, as confirmed by experiment, Newton rightly concludes that not only does the Moon gravitate towards the Earth, but the Earth also gravitates towards the Moon, and that both are in motion about a common centre of gravity, while this centre revolves continuously about the centre of gravity of the whole system. [a]

Very accurate experiments show gravity to be proportional to the quantity of matter of a solid body, other things being equal, and the same is confirmed from the calculation of the gravity of celestial bodies; but it is established by the celebrated principle and other arguments that gravity is also proportional to the material of the body towards which it is directed. The relation of other forces which predominate in nature is similar. For example, rays of light are more refracted, other things being equal, the denser the bodies which they enter. The parts of the Earth gravitate mutually towards each other and not towards that contrived point which we call the centre of the Earth; not only is this in the closest agreement with the scheme and analogy of nature, but it is also confirmed in the most beautiful way by the very accurate experiments which some most distinguished men from the Royal Academy of Paris have recently conducted in the northern part of Europe. The cause of gravity (whatever it may be precisely) dominates widely; and since it is different at different distances, it is not surprising that its force depends also on the magnitude of that body towards which it drives others. We acknowledge that this force is attributed improperly to a central body; indeed it is convenient for the sake of brevity to say so, but it is to be understood in the common sense and not in the philosophical sense.

We only touch upon these things briefly here. After Newton had determined the force of the Sun for disturbing the waters from the difference of the diameter of the equator and the axis of the Earth (which he had investigated by a certain approximation of his) following the golden rule he briefly sought the rise of the water resulting from the force of the Sun. But, whatever the

---

[a] It may be conjectured that some variation in the obliquity of the Ecliptic arises from the motion of the Sun about the centre of the system; there is discussion about this among the astronomers: it will be evidence that this is the cause of the phenomenon, if it is established that a similar variation occurs in the motion of Jupiter, the largest of the planets.

elevation of the water which is produced in this way, it differs a little from the true value, since nevertheless these problems are of a different type, the first of which depends on the quadrature of the circle, while the latter depends on that of the hyperbola or the logarithm, as we will see later; and there may be reason for doubting whether *a priori* such a short transit is entirely appropriate for the determination of the latter elevation, or also whether the method by which he had determined the figure of the Earth is sufficiently accurate; and since very subtle forces, which produce no other perceptible effects, cause the motion of the sea, so that the slightest things can be of some importance in this inquiry, I have consequently come to the conclusion that I will have done something worthwhile if I make accessible some way by which calculation in these problems can be undertaken very accurately from genuine principles.

(p.138)   First of all, a few things have to be repeated from Newton; thereafter we will follow a different path. Let L be the Moon, T the centre of the Earth, Bb the plane perpendicular to the line LT, and P any particle of the Earth; and let PM be perpendicular to the plane Bb. Let LT represent the mean gravity towards the Moon of the Earth or of the particle positioned at the centre T, let LK to LT be taken as LT^2 is to LP^2, and the line LK will be the measure of the gravity of the particle P towards the Moon. Let KG be drawn parallel to the line PT, and let it meet LT, produced if necessary, in G, and the force LK will be resolved into forces KG and LG, the first of which pushes the particle P towards the centre of the Earth and is almost equal to PT itself; the part TL of the latter, being common to all particles and always parallel to it, does not affect the motion of the water; but the other part TG is approximately equal to $3PM$:*

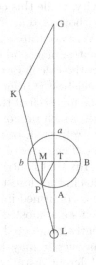

From
MacLaurin's Fig. 1

First of all, therefore, it is required to investigate what may be the figure of a fluid Earth whose particles gravitate mutually towards each other with forces decreasing in the inverse ratio of the distances squared, and which are acted upon at the same time by two external forces, one of which is directed towards the centre T, and is always as PT, the distance of the particle from the centre, while the other acts in a line parallel to TL and is to the first as $3PM$ to PT. Now we will show in the following Section that, if the Earth is supposed to be uniformly dense, the figure of this fluid is exactly a spheroid which is generated by the revolution of an ellipse about its transverse axis;

---

* This force is a little greater if the particle P is in the part of the Earth turned towards the Moon, lesser if in the part turned away from the Moon, whence it is rightly considered to be equal to $3PM$.

and hence we will endeavour to deduce the calculation of the motion of the sea from the celestial motions.

Now it is to be noted that other causes act together with the unequal gravity of the parts of the Earth towards the Moon and the Sun to produce the motion of the sea. The daily motion of the Earth about its axis seems to influence the tide of the sea in various ways, other than that noted by Newton, in which the tide is delayed to the second or third lunar hour. (1) The tide may be a little greater on account of the centrifugal force and the spheroidal figure resulting from the motion of the Earth, since this force turns out a little greater in the higher parts of the sea than in the lower lying parts. (2) Since the tide of the sea is carried either from the meridian towards the north, or in the opposite direction from the north towards the meridian, it takes place in waters, which are revolved with a different velocity about the axis of the Earth, and hence, necessarily, new motions are set up, as we will describe later. Furthermore, according to the theory of gravity, the force by which the particles of the sea are driven towards the solid Earth (which is far denser than the water) exceeds the force by which they are pushed towards the water. These forces are very small indeed; moreover, since the forces with which the Moon and the Sun act on the waters produce no perceptible effects in experiments with pendulums and in statics, yet they generate such great motions in the waters of the Ocean, it may be suspected that such small forces combine together to increase the motions of the water in some respect.

SECTION III.

Concerning the Figure which a uniformly dense fluid Earth would take on from the unequal gravity of its particles towards the Moon or the Sun.

Now that we have described the phenomena of the tide of the sea and the general principles from which it seems appropriate to seek the theory of the most celebrated phenomenon, we now pass to the determination of the figure which the fluid Earth would take on when disturbed by the forces of the Moon or the Sun described above; but there are certain Lemmas to be set out in advance, by means of which this otherwise very difficult inquiry can be made easily.

(p.141)   ### LEMMA I.

Let $ABab$ be an ellipse, C its centre, HI any diameter, and Mm the ordinate to the diameter HI at the point u; from H and m let lines HP and mx be drawn parallel to any two conjugate diameters and meeting each other in q; let qu and PM be joined, and these lines will be parallel to each other.

Let the line HP meet the ordinate Mm in z and the line MQ (which is parallel to mq) in Q. Let CG, CA and CB be semidiameters which are parallel to the lines Mm, mx and HP, respectively. Let GE be drawn parallel to CB and let it be produced until it meets the semidiameter CI in g. From the nature of the ellipse, the rectangle $Mz \times zm : Hz \times zP :: CG^2 : CB^2$; and because of the parallel lines CG and Mm, there will be $qz : zm :: GE : CG$. Hence $Mz \times qz : Hz \times zP :: CG \times GE : CB^2$. But $Hz \times zP : zu \times zP :: Hz : zu :: Gg : CG$. Consequently, by equality $Mz \times zq : zu \times zP :: Gg \times GE : CB^2$. Moreover, the rectangle formed by Gg and GE is equal to the square of the semidiameter CB by a known property of the ellipse, since CI is conjugate to the semidiameter CG, and CB to CA. Thus $Mz \times zq = zu \times zP$, and $zq : zu :: zP : zM$, and so qu is parallel to the line PM. Q.E.D.

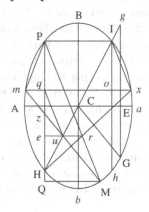

Diagram for Lemma I
MacLaurin's Fig. 2

COR. 1. The line PQ is divided harmonically in q and z, or $PQ : Pq ::$ $Qz : qz$. For if ue is drawn parallel to mx, and meets the line HP in e, then there will be

$$Pz : qz :: PM : qu \text{ (because of the parallels } PM, qu) :: PQ : qe.$$

Hence
$$Pq : qz :: Pe : qe :: qe : ez ::$$
$$Pe + qe : qe + ez :: \text{ (since } Qe, eq \text{ are equal)} \ PQ : Qz.$$

COR. 2. Let the line mx meet the ellipse in x, and let Hx be joined, which meets the line PM in r; the join ur will be parallel to mx. For if Ih is parallel to the line HP and meets mx in o, then ox will be equal to the line qm and $Io : ox :: Pq : qm :: PQ : QM$; and so Ix will be parallel to PM. But since IH is a diameter of the ellipse and the lines Ix, Hx have been drawn from the extremities of the diameter IH to the point x located on the ellipse, these two lines will be parallel to two conjugate diameters by the nature of the ellipse. Consequently, since from the points H and M, respectively, the two lines Hx and PM have been drawn parallel to the two conjugate diameters, and meet each other in r, the join ur will be parallel to the line xm by this Lemma.

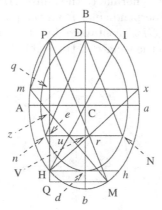

Diagram for Lemma I Cor. 3
MacLaurin's Fig. 3

COR. 3. Let the line HP now be parallel to the axis of the ellipse, and the angle HPM will be equal to the angle HPm, since $QM : qm :: Qz : qz :: PQ : Pq$ by Cor. 1. Next let Hh and PI be drawn parallel to the other axis Aa and let them meet the axis Bb in D and d; on the axis Dd let the ellipse be described which is similar to the ellipse $ABab$ and similarly positioned, and let the line ur produced meet it in N and n; let ur meet the axis Dd in V, and VN or Vn will be equal to the line er, and if Dn, DN are joined, these lines will be parallel to the lines PM, Pm, respectively. For $Pe : er :: Pq : qm$ and $He : er :: Hq : qx$, hence

$$He \times Pe : er^2 :: Hq \times qP : mq \times qx :: CB^2 : CA^2.$$

But the rectangle $DV \times Vd : VN^2 :: CB^2 : CA^2$; $dV = He$, $DV = Pe$, and so $DV \times Vd = He \times Pe$, hence $VN^2 = er^2$, and $VN = er$, PM is parallel to the line DN and Pm to the line Dn.

COR. 4. Hence we have the following converse where Nn is an ordinate from the interior ellipse to the axis Dd and DP, which is perpendicular to the axis Dd, meets the exterior ellipse in P; let DN and Dn be joined and let PM, Pm, which are parallel to these lines, meet the exterior ellipse in M and m; let PH be drawn parallel to the axis Dd, and let MQ and mq be perpendicular to PH; then $PQ + Pq$ (or $2Pe$) will be equal to $2DV$ if the points Q and q lie on the same side of the point P, and $PQ - Pq = 2DV$ when Q and q are on opposite sides of the point P.

(p.145)

LEMMA II.

Let the line PL which is normal to the ellipse $ABab$ at P meet the axis Bb in L, and let LZ be the perpendicular from the point L to the semidiameter CP, and the rectangle CPZ contained by the semidiameter CP and the intercept PZ is equal to the square of the semiaxis CA.

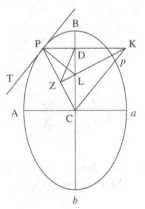

Diagram for Lemma II
MacLaurin's Fig. 4

Let Cp be a semidiameter conjugate to CP, let PD be drawn perpendicular to the axis Bb and let it be produced until it meets the semidiameter Cp in K, let KZ be joined, and let PT be the tangent to the ellipse at the point P. Because of the right angles LDP, LZP, the circle LPZ will pass through the four points L, D, P, and Z, and will touch the line PT at P, and so the angle PDZ will be equal to the angle CPT or PCK. Thus a circle will pass through the four points C, K, D and Z; the angle CZK will be equal to the right angle CDK, KZ will pass through the point L and from the nature of the circle $CP \times PZ = DP \times PK = CA^2$. Q.E.D. [a]

(p.146)

LEMMA III.

Let us assume that particles of bodies gravitate towards each other with forces which decrease in the inverse ratio squared of the distances between them, and let $PAEa$, $PBFb$ be similar pyramids or cones made up of homogeneous material of this type, and the gravity of the particle P into the solid $PAEa$ to the gravity of the same particle into the solid $PBFb$ will be as PA to PB, or as any corresponding sides of these solids.

For the gravity of particle P into any surface $AEaA$ concentric with the point P is as this surface directly and the square of the radius PA inversely,

---

[a] The properties demonstrated in this and the preceding Lemmas are likewise easily carried over to the hyperbola.

and so it is always the same at any distance PA. Consequently, the gravity of the particle P towards the whole solid $PAEa$ will be to the gravity of the same particle towards the whole solid $PBFb$ as PA to PB.

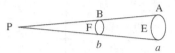

Diagram for Lemma III
MacLaurin's Fig. 5

COR. 1. Hence the gravities by which particles which are similarly situated with respect to similar homogeneous solids are pushed towards these solids, are as the distances of the particles from similarly situated points in these solids, or as any corresponding sides of the solids. For these solids can be resolved into similar cones or pyramids, or similar parts of these, which will have vertices at the gravitating particles.

COR. 2. Hence it also follows easily that, if an elliptic annulus which is bounded by similar figures $ABab$, $DndN$ is revolved about either of its axes, the gravity towards this solid of a particle located inside the solid so generated, or placed on its interior surface, vanishes; for if any line meets these similar and similarly positioned ellipses, the extreme segments of the line which are cut off by the ellipses will always be equal (as is easily shown from the nature of these figures) and so in this case equal and opposite forces will destroy each other. Indeed it follows from this that, if $ABab$ is the spheroid generated by the motion of an ellipse about either of its axes, and if B and D are any particles located in the same semidiameter, the gravity of the particle B towards the spheroid will be to the gravity of the particle D as the distance CB to the distance CD, by the preceding Corollary.

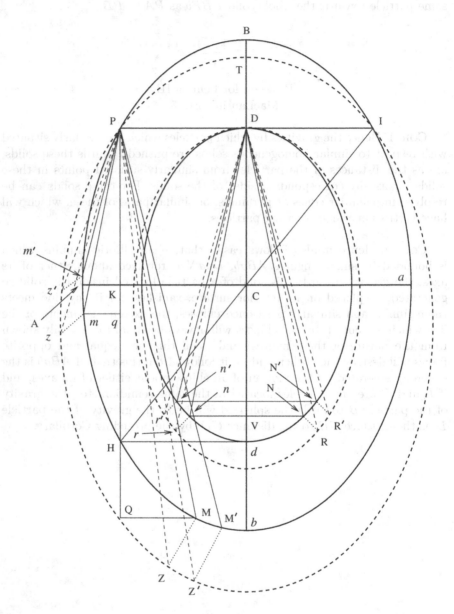

Diagram for Lemma 4
MacLaurin's Fig. 6

LEMMA IV.

Let $ABab$ be the spheroid generated by the motion of the semiellipse ABa about the axis Aa, and for any particle P on the surface of the solid, let PK be normal to the axis at K; and let PD, which is parallel to the axis, meet the plane Bb (which is supposed normal to the axis) in D. Let the force with which the particle P gravitates towards the spheroid be resolved into two forces, one parallel to the axis, the other perpendicular to it, and the former will be equal to the force with which the particle K located on the axis tends to the centre of the solid, while the latter will be equal to the force by which the particle D is pushed towards the same centre.

Let PK be produced until it meets the generating ellipse again in H, let Hd be drawn parallel to the axis Aa and let it meet the axis Bb in d; let us imagine described on the axis Dd the solid $DndN$ which is similar to $BAba$ and similarly situated. Sections of these solids cut off by the same plane will always be similar, similarly situated ellipses, as is known and easily shown. Therefore let $BAba$, $DndN$ be figures of this type cut off from these similar solids by the plane $PAbIBP$, which is assumed always to pass through the given line PDI. Let the plane $PzZIT$ contain an exceedingly small angle with the first plane and let it make similar, similarly positioned sections $PzZIT$, $DrRD$ in the surfaces of the aforesaid solids. These things having been put in place, we will show first of all that the force by which the particle P is pushed towards the two parts which are contained by the planes PbI, PZI and by the planes PBI, PTI, if it is reduced to the direction PK, will be equal to the force by which the particle D is pushed towards the part bounded by the planes $DnND$, $DrRD$.

For let Nn, $N'n'$ be two ordinates from the internal ellipse to the axis Dd; let [a] PM, Pm, PM' and Pm' be parallel to the lines DN, Dn, DN' and Dn', respectively; furthermore, let the planes DNR, $DN'R'$, Dnr, $Dn'r'$, PMZ, $PM'Z'$, Pmz, $Pm'z'$ be perpendicular to the plane $PbIB$, and let them meet the other plane $PzZIT$ in the lines DR, DR', Dr, Dr', PZ, PZ', Pz, Pz', respectively. These things having been put in place, since the angles NDN' and MPM', nDn' and mPm' are always put equal, and the lines PM and DN, Pm and Dn are always equally inclined to PI, the common intersection of the planes, if the angle NDN' and the inclination of the planes $PbIB$, $PZIT$ to each other be supposed to decrease continuously until they vanish, the gravities of the particle D into the pyramids $DNN'R'R$, $Dnn'r'r$ and of the particle P into the pyramids $PMM'Z'Z$, $Pmm'z'z$ will be in the limit as the lines DN, Dn, PM and Pm, respectively, by Lemma 3. And the same forces determined along the lines perpendicular to the axis Aa will be

---

[a] In drawing this figure we have not drawn the lines NR, $N'R'$, etc., according to the rules of perspective, but in that manner by which they can be most easily distinguished one from another.

as the lines DV, DV, PQ, Pq, respectively. Hence, since $PQ \mp Pq = 2DV$ by Cor. 4 of Lemma 1, it follows that the force by which the particle P is pushed towards the axis Aa by its gravity into the pyramids $PMM'Z'Z$, $Pmm'z'z$, is equal to the force by which the particle D is pushed by its gravity towards the pyramids $DNN'R'R$, $Dnn'r'r$. Consequently, if the planes DNR, PMZ are moved everywhere about the points D and P, while always parallel to each other and perpendicular to the plane $PbIB$ (the lines DN, PM of course going forward always in the plane $PbIB$, and the lines DR, Pz in the plane $PZIT$), the forces by which the particle P is pushed towards the axis by its gravity into the parts thus described by the motion of the planes PMZ, Pmz, will always be equal to the forces by which the particle D is pushed towards the same axis by its gravity into the parts described by the motion of the planes DNR, Dnr; hence it follows that the particle P is pushed by the same force along the line PK, by its gravity into the parts contained by the planes PbI, PzI, and by the planes PBI, PTI, as that with which the particle D tends towards the parts bounded by the planes $DnND$, $DrRD$. Thus, since these forces may also be calculated to be equal along lines perpendicular to the axis of the whole solid, and the ratio of the forces by which the particles P and D are pushed towards any other parts cut similarly from the solids is the same, it follows that the particle P is pushed towards the axis by its gravity into the exterior solid equally as the particle D is pushed likewise by its gravity into the interior solid, or also into the exterior solid, since these forces are the same by Cor. 2 of Lemma 3.

It is established in an entirely similar way that the force by which the particle P is pushed along the line parallel to the axis is equal to the force by which the particle K located in the axis is pushed towards the centre of the solid.

Cor. 1. Therefore any particles of a spheroid which are at equal distances from the axis or the equator of the solid are pushed equally towards the axis or the equator. And the forces by which any particles are pushed towards the axis are as their distances from the axis, and the forces by which they are pushed towards the plane of the equator are to each other as their distances from this plane.

Cor. 2. Let $\mathcal{A}$ represent the force with which the spheroid attracts a particle located on the axis at the extremity A and $\mathcal{B}$ the force with which the same solid attracts the particle B positioned between A and a on the circumference of the middle circle; let KR to KC be taken as $\frac{A}{CA}$ is to $\frac{B}{CB}$ and let PR be joined; and the particle P will tend towards the spheroid along the line PR, with a force which is always proportional to this line. For the force by which the particle D is pushed towards the centre of the solid, is to $\mathcal{B}$ as CD to CB, by Cor. 2 of Lemma 3. Similarly the force by which the particle K is pushed towards the centre of the solid is to $\mathcal{A}$ as CK to CA. Consequently, by Lemma 4 the force by which the particle P is pushed along the line PK normal to the axis is to the force by which it is pushed along the line PD parallel to the axis, as $\frac{PK \times B}{CB}$ to $\frac{CK \times A}{CA}$; and so this is as $PK \times KC$ to $CK \times KR$, i.e., as PK to KR by construction. Consequently the particle P is pushed along the line PR when these forces have been put together and the composite force is to B as PR to BC. In fact the forces A and B can be computed in this way, as we will show later.

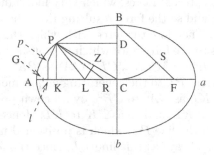

Diagram for Lemma IV Cor. 2 and Prop. I
MacLaurin's Fig. 7

(p.157)

PROPOSITION I.

FUNDAMENTAL THEOREM.

Let the spheroid $ABab$ be composed of fluid material, whose particles are attracted towards each other mutually by forces which decrease in the inverse ratio squared of the distances; and at the same time let two external forces act on the individual particles of the fluid, one of which tends towards the centre of the spheroid, and is always proportional to the distances of the particles from this centre, while the other acts along lines parallel to the axis of the solid, and is always proportional to the distances of the particles from the plane Bb normal to the axis; and if the semiaxes CA, CB of the generating ellipse are inversely proportional to the whole forces, which act on equal particles located at the extreme points A, B of the axes, the entire fluid will be in equilibrium.

In order that this our first Proposition may be demonstrated with the greatest clarity, we will show first of all that the force composed of the gravity of any particle P and the two external forces always acts in the line PL which is always normal to the surface of the spheroid. (2) The fluid in any line PC drawn from the surface to the centre is of the same weight everywhere. (3) The fluid in any canals drawn from the surface to any given particle within the solid always attracts that particle with the same force.

1. Let the total forces which act on the particles A and B be called M and N, which by hypothesis are in the ratio of the axes CB and CA. Let the first external force which acts in the line PC be resolved into two forces, one

parallel to the axis, the other perpendicular to it; and these forces will always be as the lines PK and KC. Hence, since the force with which the gravity of the particle P pushes it along the line PK is also as PK, by the previous Lemma, it follows that the total force with which the particle P is pushed along the line PK is to N as PK to CB. Three forces act on the particle P along the line PD which is parallel to the axis, namely, the gravity of the particle and the two external forces, which vary individually in proportion to the line PD or KC; and so the force resulting from these three forces will be to M as CK to CA. Therefore the force with which the particle P is pushed along the line PK is to the force with which it is pushed along the line PD as $\frac{N \times PK}{CB}$ to $\frac{M \times KC}{CA}$ or (since $M : N :: CB : CA$) as $PK \times CA^2$ to $CK \times CB^2$, i.e., (since if the normal PL to the generating ellipse meets the axis Aa in L, then KC will be to KL as CA^2 to CB^2, by a known property of the ellipse) as $PK \times KC$ to $KC \times KL$, and so as PK to KL. Hence the composite force pushes the particle in the line PL, which is positioned normal to the surface of the fluid; and it is always as this line PL, since the forces along the lines PK are always as PK.

2. Let LZ be perpendicular to the semidiameter CP, and the force with which the particle P is pushed towards the centre will be as the line PZ by the common principles of Mechanics, and the weight of fluid in the line PC will be as the rectangle $CP \times PZ$, which is always equal to the square of the semiaxis CB by Lemma II. Therefore the centre is pushed equally from all directions, and the fluid is in equilibrium at C.

3. Let p be any particle located anywhere in the solid, and Pp any line drawn from the surface to the particle p; let PK, pl be perpendicular to the axis Aa; and, by an easy calculation which I omit for the sake of brevity, the force by which the particle p is pushed by the weight of fluid in the line Pp in the direction of this line will be found to be equal to

$$\frac{N}{2CB} \times \overline{PK^2 - pl^2} - \frac{M}{2CA} \times \overline{Cl^2 - CK^2} = \text{ (since } M : N :: CB : CA)$$

$$\frac{M \times CA^2 \times PK^2 + M \times CK^2 \times CB^2 - M \times CA^2 \times pl^2 - M \times CB^2 \times Cl^2}{2CB^2 \times CA}$$

$$= \frac{M \times CA^2 - M \times CG^2}{2CA},$$

(since $PK^2 : CA^2 - CK^2 :: CB^2 : CA^2$, and if CG is the semiaxis of the ellipse drawn through p similar to the ellipse $ABab$ and similarly situated, then $pl^2 : CG^2 - Cl^2 :: CB^2 : CA^2$) and so, since these quantites do not depend on the location of the point P, this force is always the same, if the position of the particle p is given; thus, since p is pushed equally from all directions, the fluid will be in equilibrium everywhere.

COR. 1. As in Cor. 2 of Lemma IV let $\mathcal{A}$ be the force of gravity into the spheroid at the place A, let $\mathcal{B}$ the the force of gravity into the spheroid at the place B, let V be the force KG (Fig. 1), shown in the upper section in its mean value, by which the Moon or the Sun depresses the water of the spheroid at the distance d which is put in the middle between CA and CB. Let $CA = a$, $CB = b$, and the force N, with which the particle B is pushed towards C, will be equal to $\mathcal{B} + \frac{bV}{d}$, and

$$M = \mathcal{A} + \frac{aV}{d} - \frac{3aV}{d} = \mathcal{A} - \frac{2aV}{d}.$$

Hence by this Proposition, if $a : b :: \mathcal{B} + \frac{bV}{d} : \mathcal{A} - \frac{2aV}{d}$, the fluid will be in equilibrium. And hence if $\mathcal{A}$, $\mathcal{B}$ and V are given, the type of the figure will become known in terms of a and b. There is

$$\mathcal{A}a - \mathcal{B}b = \frac{2a^2V}{d} + \frac{b^2V}{d}.$$

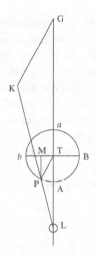

From MacLaurin's Fig. 1
(for the Corollaries)

Cor. 2. Since the force V (let it arise either from the unequal gravity of particles towards the Moon, or towards the Sun) is extremely small relative to the forces $\mathcal{A}$ and $\mathcal{B}$, and the difference between a and b is quite small, put $a = d + x$ and $b = d - x$, and there will be

$$\mathcal{B}d - \mathcal{B}x + V \times \frac{(d - x)^2}{d} = \mathcal{A}d + \mathcal{A}x - 2V \times \frac{(d + x)^2}{d},$$

and, when terms have been neglected in which x^2 is found,

$$\mathcal{B}d - \mathcal{B}x + Vd - 2Vx = \mathcal{A}d + \mathcal{A}x - 2Vd - 4Vx,$$

hence

$$Bd - Ad + 3Vd = Ax + Bx - 2Vx \, ;$$

and so

$$x : d :: B - A + 3V : B + A - 2V \, ;$$

and the difference of the height of the water at A and B (or $2x$) is to the mean semidiameter d as $2B - 2A + 6V$ to $B + A - 2V$, or approximately as $B - A + 3V$ to the mean gravity towards the spheroid.

COR. 3. In the preceding Corollaries we have supposed $d = \frac{1}{2}CA + \frac{1}{2}CB$; but if d denotes some other distance where the force KG (Fig. 1) is put equal to V, and there is $e = \frac{1}{2}CA + \frac{1}{2}CB$, then there will be

$$x : e :: B - A + \frac{3eV}{d} : B + A - \frac{2eV}{d} \, .$$

COR. 4. By the force V in these Corollaries we have understood the force of either the Sun or the Moon, and we have considered the figure which a fluid homogeneous Earth would take on if these forces act separately on it. Now let the Moon be in conjunction with the Sun or opposed to it, and let them act together on the Earth. In this case the forces of the luminaries conspire together to raise the water at A and a, and to depress it at B and b, and they observe the same laws everywhere. Hence in this case also the fluid will be in equilibrium, if the total force which acts at the place A is to the total force which acts at the place B as CB to CA; and so if V now denotes the sum of the forces with which the Sun and the Moon depress the water in the lines Tb, TB (Fig. 1) at the mean distance, the fluid will be in equilibrium if

$$b : a :: A - \frac{2aV}{d} : B + \frac{bV}{d} \, ,$$

or x is to d as $B - A + 3V$ is to $B + A - 2V$ approximately, as before.

COR. 5. Now let the Moon be in the line Aa and the Sun in the line Bb; and since the force of the Moon is more powerful, let the transverse axis of the generating figure pass through the Moon and the conjugate axis through the Sun; and if the total force which acts at the place A is to the total force which acts at the place B as CB to CA, the fluid spheroid will be in equilibrium also in this case. Let s be the force with which the Sun depresses the water in the lines TA, Ta at the mean distance from the centre C, l the force with which the Moon depresses the water in the lines TB, Tb at equal distance; and the total force which acts at the place A will be equal to

$$A - \frac{2al}{d} + \frac{as}{d} \, ,$$

and the total force which acts at the place B will be equal to

$$\mathcal{B} + \frac{bl}{d} - \frac{2bs}{d} \, .$$

Hence we put together as in Cor. 2

$$x : d :: \mathcal{B} - \mathcal{A} + 3l - 3s : \mathcal{B} + \mathcal{A} - 2l - 2s ::$$

(if $l - s$ is now called V) $\mathcal{B} - \mathcal{A} + 3V : \mathcal{B} + \mathcal{A} - 2V,$

as before.

SCHOL. It is shown in exactly the same way that if $BabA$ is an oblate fluid spheroid generated by the motion of the semiellipse BAb about its minor axis Bb and this spheroid is turned about the same axis by a motion such that the gravity towards this spheroid at the pole B is to the excess by which the gravity at the place A exceeds the centrifugal force at A arising from the motion of the spheroid about the axis as CA to CB, then the fluid will be in equilibrium everywhere. Hence it follows that the figure of the Earth, however it is altered by the centrifugal force arising from the daily motion, is an oblate spheroid of the type generated by the motion of a semiellipse BAb about the minor axis (if the material of the Earth is assumed to be equally dense), the semidiameter of the equator is to the semiaxis as the gravity at the pole on the Earth is to the excess of gravity over the centrifugal force at the equator, a body at any place P tends towards the Earth with a force which is always as the line PL normal to the generating ellipse and meeting the major axis in L, and finally the measure of a degree in the meridian is always as the cube of the same line PL. These things are all demonstrated precisely from this Proposition; since they are of special use in the inquiry into the figure of the Earth, it is only appropriate to mention them here in passing.

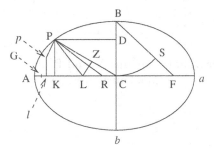

Diagram for the Scholium
MacLaurin's Fig. 7

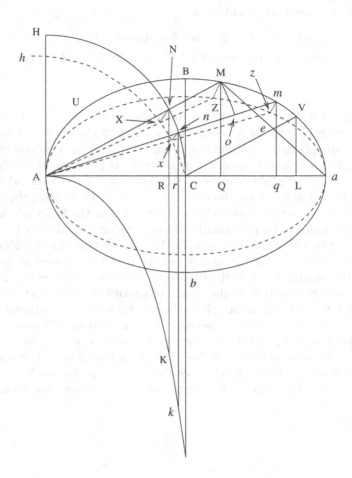

Diagram for Lemma V and Proposition II
MacLaurin's Fig. 8

LEMMA V.

Let there be any figure ABa: let the circle CNH be described with centre A and any given radius AC; from A let any line AM be drawn meeting the figure ABa in M and the circle in N; let MQ and NR be perpendiculars to the given axis Aa, let KR be always equal to the abscissa AQ, and the force with which the particle A is attracted towards the solid generated by the motion of the figure ABa about the axis Aa will be as the area which the ordinate KR generates directly and the radius AC inversely.

Let another line drawn from A meet the figure in m and the circle in n, and let mq and nr be perpendiculars to the axis Aa. Let $AZza$ be another section through the axis of the solid which the planes AMz, Amz, normal to AMa, meet in the lines AZ, Az, which cut the circle drawn in the plane $AZza$ with radius AC in X and x; finally let the circular arc Mo drawn with centre A meet Am in o. These things having been put in place, let the angle contained by the planes AMa, AZa be decreased, and at the same time the angle MAm, until they vanish and the ultimate ratio of the force with which the particle A tends towards the pyramid $AMZzm$ to the force with which it is attracted towards the pyramid $ANXxn$ will be the ratio of the line AM to the line AN, or AQ to AR, by Lemma III. The force of this pyramid is as the force of the surface $NXxn$ multiplied by the line AN, and so as $\frac{NX \times Nn}{AN^2} \times AN = \frac{NX \times Nn}{AN}$, or as $\frac{NR \times Nn}{AN}$ (since NX is as NR) i.e., as Rr; and the force of the same reduced in the direction of the axis will be as $Rr \times \frac{AR}{AN}$; consequently, the force of the pyramid $AMZzm$ reduced in the same direction will be as $Rr \times \frac{AQ}{AC} = \frac{Rr \times KR}{AC}$. Therefore the force with which the particle A is attracted towards the part of the solid contained by the planes AMa, Aza is as the area which the ordinate KR generates directly and the radius AC inversely; and since the solid is round, having been generated by the motion of the figure about the axis Aa, the ratio of the force with which the particle is attracted towards the whole solid will be the same.

COR. The force with which the particle A is attracted into the solid is to the force with which it is attracted towards the sphere drawn on the diameter Aa as the area which the ordinate KR generates to $\frac{2}{3}CA^2$. For if AMa is a circle, there will be AQ to Aa as AQ^2 to AM^2, or AR^2 to AN^2. Hence, in this case there will be $KR = \frac{2AR^2}{AC}$, and the area ARK (which the ordinate KR generates) $= \frac{2AR^3}{3AC}$, and so the whole area generated by the motion of the ordinate RK will be $\frac{2}{3}CA^2$.

PROPOSITION II.

PROBLEM.

To find the gravity towards an oblong spheroid of a particle A situated at an extremity of the transverse axis.

Other things remaining as in the preceding Lemma, let AMa be an ellipse, Aa its transverse axis, C its centre, Bb its conjugate axis, and F a focus; let any line AM be drawn from A meeting the ellipse in M, and let the line CV parallel to it meet the ellipse in V; from this point let the ordinate VL be drawn to the axis, let the join aM meet the line CV in e, and there will be $AM = 2\,Ce$: and since

$$AQ : CL :: AM\,(2\,Ce) : CV :: 2\,CL : Ca\,,$$

the quantities $\frac{1}{2}AQ$, CL and CA will be in continued proportion. Let $CA = a$, $CB = b$, $CF = c$, $AR = x$, $CL = l$, and since $AR^2 : NR^2 :: CL^2 : VL^2$ there will be

$$x^2 : a^2 - x^2 :: l^2 : \overline{a^2 - l^2} \times \frac{b^2}{a^2}\,;$$

and so

$$l^2 = \frac{a^2b^2x^2}{a^4 - c^2x^2} \quad \text{and } AQ \text{ or } KR = \frac{2l^2}{a} = \frac{2ab^2x^2}{a^4 - c^2x^2}\,,$$

and

$$\text{area } ARK = \int \frac{2ab^2x^2\,dx}{a^4 - c^2x^2} = (\text{if } z : x :: c : a)\ \int \frac{2a^2b^2}{c^3} \times \frac{z^2\,dz}{a^2 - z^2}\,.$$

Consequently, let a be the quantity whose logarithm vanishes, or the modulus of the logarithmic system, ℓ the logarithm of the quantity $a\sqrt{\frac{a+z}{a-z}}$, and there will be $ARK = \frac{2a^2b^2}{c^3} \times \overline{\ell - z}$. Hence the force with which the particle A gravitates towards the solid generated by the motion of the elliptic segment $AUMA$ about the axis Aa, will be to the force with which the same particle gravitates towards the solid generated by the motion about the same axis of the circular segment cut off from the circle drawn on the diameter Aa by the same line AM as

$$\frac{2a^2b^2}{c^3} \times \overline{\ell - z} \quad \text{to} \quad \frac{2x^3}{3a}\,;$$

and if $\mathcal{L}$ is the logarithm of the quantity $a\sqrt{\frac{a+c}{a-c}}$ (or $\frac{a}{b} \times \overline{a+c}$), the force with which the particle A tends towards the whole spheroid will be to the force with which it tends towards the whole sphere as $3b^2 \times \overline{\mathcal{L} - c}$ to c^3.

SCHOL. In the same way the gravity towards an oblate spheroid of a particle situated at the pole is found by investigating the area whose ordinate is $\frac{2b^2a^2}{c^3} \times \frac{z^2}{b^2+z^2}$. Let $BAba$ be the oblate spheroid generated by the motion of an ellipse BAb about its minor axis; with centre B and radius BC let the circular arc CS be described, meeting the line BF in S, and the gravity into this spheroid at the pole B will be to the gravity at the same place towards the sphere drawn on the diameter Bb as $3CA^2 \times \overline{CF - CS}$ to CF^3. However, the method by which the gravity towards an oblong or an oblate spheroid of a particle situated at the equator may be computed is less obvious, but it comes out easily by use of the following Lemma.

(p.172)

<h1 style="text-align:center">LEMMA VI.</h1>

Let two planes $BMbaB$, $BZgeB$ cut each other in the line HBh, which is a common tangent of the figures, and let them take off from the solid the part $BMbaBZgeB$; let the semicircles HCh, Hch be the sections of these planes and the surface of the sphere described with centre B and radius BC. From the point B let any line BM be drawn in the first plane meeting the figure $BMba$ in M and the semicircle HCh in N; and let MQ and NR be perpendicular to Hh and let the ordinate KR be always equal to the line MQ. These things having been put in place, if the angle CBc contained by these planes is decreased indefinitely, the gravity of the particle B towards the part $BMbaBZgeB$ will be ultimately to the gravity of the same particle towards the part of the sphere contained by the semicircles HBh, Hch as the area $HKdh$ generated by the motion of the ordinate KR to the semicircle HCh.

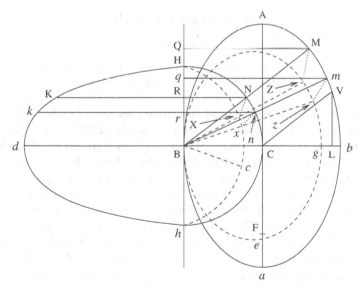

Diagram for Lemma VI and Prop. III
MacLaurin's Fig. 9

Let m be a point in the figure BMb very close to M and let Bm be joined, which meets the circle HCh in n; and let nr be perpendicular to Hh.

In addition, let the planes BMZ, Bmz be perpendicular to the plane $BMba$, and let them cut the other plane $BZge$ in the lines BZ, Bz, which meet the circumference Hch in X and x. These things having been put in place, the force with which the particle B gravitates into the pyramid $BMZzm$ will be ultimately to the force with which the same particle gravitates into the pyramid $BNXxn$ as the line BM to BN, or MQ to NR by Lemma III. But the gravity into the latter pyramid is as $\frac{NX \times Nn}{BN^2} \times BN$, or (since NX is as NR) as $\frac{NR \times Nn}{BC}$, i.e., as Rr; and this gravity acts along the line Bb with a force which is as $\frac{Rr \times RN}{BC}$; hence the gravity into the pyramid $BMZzm$ acts along the line Bb with a force which is as $\frac{Rr \times MQ}{BC}$, or $\frac{Rr \times KR}{BC}$. Thus the ultimate ratio of the forces with which the particle B is attracted towards the whole parts of the solid and of the sphere BC is the ratio of the area $HKdh$ (which the ordinate KR generates) to the semicircle HCh.

Cor. The gravity into the part bounded by the planes $BMba$, $BZge$ is to the gravity into the spherical part contained by the circles described on the diameters Bb, Bg as the area $HKdh$ to $\frac{8}{3}CB^2$. For let $BMbB$ be a circle, and there will be MQ to Bb as RN^2 to BC^2, and

$$KR = \frac{2\,RN^2}{CB} = \frac{2\,BC^2 - 2\,BR^2}{CB} \quad \text{and} \quad \text{area}\,HKdB = \frac{4}{3}CB^2,$$

and so the total area $HKdh = \frac{8}{3}CB^2$.

(p.174)

PROPOSITION III.

Problem.

To find the gravity towards an oblong spheroid
of a particle situated at the equator.

By the equator we understand the circle generated by the conjugate axis while the figure is rotated about the other axis. Let $BMba$ in the figure of the preceding Lemma represent any section of the spheroid normal to the plane of the equator, and this figure will always be similar to the section of the solid through the poles, or to the figure by whose rotation we suppose the solid to be generated. For the sake of brevity I omit the demonstration of this, which is easy and has been treated by others. Therefore let CA be a transverse semiaxis of this section, CB a conjugate semiaxis and F a focus; let $CB = b$, $CA = a$, $CF = c$, $BR = x$; let CV be a semidiameter parallel to the line BM, VL the ordinate to the axis Bb and $CL = l$. Then $CB : CL :: CL : \frac{1}{2}MQ$ as in the preceding Proposition, and $MQ = \frac{2l^2}{b}$. But $NR^2 : BR^2 :: CL^2 : VL^2$, i.e.

$$b^2 - x^2 : x^2 :: l^2 : \overline{b^2 - l^2} \times \frac{a^2}{b^2}, \quad \text{or} \quad a^2 - \frac{a^2 x^2}{b^2} : x^2 :: l^2 : b^2 - l^2,$$

and

$$l^2 = \frac{a^2 b^2 \times \overline{b^2 - x^2}}{a^2 b^2 - c^2 x^2} = (\text{if } z : x :: c : b) \; \frac{b^2 a^2}{c^2} \times \frac{c^2 - z^2}{a^2 - z^2},$$

and

$$KR = MQ = \frac{2l^2}{b} = \frac{2a^2 b}{c^2} \times \frac{c^2 - z^2}{a^2 - z^2},$$

and the area $BdKR$ is equal to

$$\int \frac{2a^2 b^2 dz}{c^3} \times \frac{c^2 - z^2}{a^2 - z^2} = \frac{2a^2 b^2 z}{c^3} - \int \frac{2a^2 b^2}{c^3} \times \frac{b^2 dz}{a^2 - z^2}.$$

Therefore let ℓ (as in the previous Proposition) be the logarithm of the quantity $a\sqrt{\frac{a+z}{a-z}}$, and the area $BdKR$ will be

$$\frac{2a^2 b^2 z}{c^3} - \frac{2a^2 b^2}{c^3} \times \frac{b^2 \ell}{a^2} = \frac{2b^2}{c^3} \times \overline{a^2 z - b^2 \ell}.$$

Now let $x = b$, and so $z = c$, be substituted; and let $\mathcal{L}$ be the logarithm of the quantity $a\sqrt{\frac{a+c}{a-c}}$, as before, and the whole area $HKdh$ generated by the motion of the ordinate KR will be equal to $\frac{4b^2}{c^3} \times \overline{a^2 c - b^2 \mathcal{L}}$. Consequently the gravity of the particle B towards the part bounded by the elliptic planes $BMba$, $BZge$ will be ultimately to the gravity into the part contained by the same planes and cut off from the sphere described with centre C and radius CB as $a^2 c - b^2 \mathcal{L}$ to $\frac{2}{3} c^3$ by the Corollary to Lemma VI. Let the circle $BPpb$ be the equator of the spheroid and BP, Bp any two chords of this circle; sections of the spheroid perpendicular to the circle BPb will be ellipses similar to the section which passes through the poles of the solid, of which BP and Bp will be transverse axes; moreover, the sections by the same planes of the sphere described on the diameter Bb will be circles whose diameters will be the chords BP, Bp. Thus the ratio of the gravity of the particle B into the elliptical part and into the spherical part bounded by these planes will always be the same; and the gravity towards the whole spheroid will be to the gravity towards the sphere as $a^2 c - b^2 \mathcal{L}$ to $\frac{2}{3} c^3$, where a denotes the transverse semiaxis of the figure by whose motion the solid is generated, b the conjugate semiaxis, c the distance of the focus from the centre, and $\mathcal{L}$ the logarithm of $a\sqrt{\frac{a+c}{a-c}}$ or $a \times \frac{a+c}{b}$. Q.E.F.

COR. The ratio of the gravity towards any part of the spheroid and the part of the sphere cut off by the same plane normal to the equator and on the same side of the plane is always the same; or the gravity into the portion cut off from the spheroid by this plane is to the gravity into the whole spheroid as the gravity into the part of the sphere cut off by the same plane on the same side to the gravity into the whole sphere.

SCHOL. In the same way, if $BAba$ is the oblate spheroid generated by the motion of the figure BAb about the minor axis Bb, the gravity into this spheroid at the place A will be to the gravity at the same place towards the sphere described with centre C and radius CA as $CA^2 \times CS - CB^2 \times CF$ to $\frac{2}{3}CF^3$.

(p.177)

PROPOSITION IV.

PROBLEM.

Given the forces with which the particles of the Earth gravitate towards the Sun and the Moon, to find the figure which the Earth would take on in the syzygies or the quadratures of the Sun and Moon under the hypothesis that the Earth is made up of a homogeneous fluid and does not move around its axis.

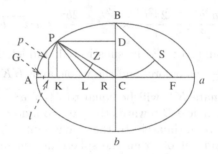

MacLaurin's Fig. 7

The gravity at the place A towards the oblong spheroid generated by the motion of the figure ABa about the transverse axis Aa is to the gravity at the same place towards the sphere described with centre C and radius CA as $3b^2 \times \overline{\mathcal{L} - c}$ to c^3 by Prop. II. Moreover, the latter gravity is to the gravity at B towards the sphere described with centre C and radius CB as CA to CB (by Cor. I of Lemma III), which is to the gravity at the place B towards the spheroid as $\frac{2}{3}c^3$ to $a^2c - b^2\mathcal{L}$ by Prop. III. Let these ratios be put together and the gravity at the place A towards the spheroid will be to the gravity at the place B towards the same body as $2ab \times \overline{\mathcal{L} - c}$ to $a^2c - b^2\mathcal{L}$. Let $\mathcal{A}$ denote the gravity at the place A, $\mathcal{B}$ the gravity at the place B, V the sum of the forces by which the luminaries, in conjunction or opposed, depress the water in the lines TB, Tb (Fig. 1) perpendicular to the line Aa which is supposed to pass through the centres of the Earth and of the luminaries, as in Cor. 4 of Prop. I, or the difference of the same forces in the quadratures of the Moon, as in Cor. 5 of the same Prop. and by the things that are demonstrated in Cor. 1 of Prop. I there will be

$$\mathcal{A}a - \mathcal{B}b = \frac{2a^2V + b^2V}{d}.$$

And so

$$\mathcal{A}a - b\mathcal{A} \times \frac{a^2c - b^2\mathcal{L}}{2ab \times \overline{\mathcal{L} - c}} = \frac{2a^2V + b^2V}{d},$$

and

$$V : \mathcal{A} :: 2a^2\mathcal{L} + b^2\mathcal{L} - 3a^2c : \frac{2a}{d} \times \overline{2a^2 + b^2} \times \overline{\mathcal{L} - c}.$$

And given the ratio of V to $\mathcal{A}$ or to $\mathcal{B}$, or $\frac{1}{2}\mathcal{A} + \frac{1}{2}\mathcal{B}$ (which can be taken for G the mean gravity at the circumference $ABab$), we will have an equation from which the type of the figure and the difference of the semiaxes or the rise of the water can be calculated.

Now the logarithm $\mathcal{L}$ of the quantity $a\sqrt{\frac{a+c}{a-c}}$ is equal to

$$c + \frac{c^3}{3a^2} + \frac{c^5}{5a^4} + \frac{c^7}{7a^6}, \&c.$$

by very well-known methods, and so

$$\mathcal{L} - c = \frac{c^3}{3c^2} + \frac{c^5}{5a^4} + \frac{c^7}{7a^6}, \&c.$$

Hence V is to $\mathcal{A}$ as[59]

$$\frac{2c^2}{15a^2} + \frac{4c^4}{35a^4} + \frac{6c^6}{63a^6}, \&c. \quad \text{to} \quad \frac{\overline{\mathcal{L} - c} \times a \times \overline{2a^2 + b^2}}{dc^3},$$

and V is to $\frac{1}{2}\mathcal{A} + \frac{1}{2}\mathcal{B}$ or G as

$$\frac{2c^2}{15a^2} + \frac{4c^4}{35a^4} + \frac{6c^6}{63a^6}, \&c. \quad \text{to} \quad \frac{2a^2 + b^2}{4bdc^3} \times \overline{2ab\mathcal{L} - b^2\mathcal{L} + a^2c - 2abc}.$$

But if V is sufficiently small in relation to the gravity G (as in the present case) the difference of the semidiameters CA, CB will be to the mean semidiameter approximately as $15V$ to $8G$, or a little more accurately, as $15V$ to $8G - 57\frac{5}{14} \times V$. For, as in Cor. 2 of Prop. I, let $a = d + x$, $b = d - x$, so that $c^2 = a^2 - b^2 = 4dx$, and there will be

$$\mathcal{A} : \mathcal{B} :: 2ab \times \overline{\mathcal{L} - c} : a^2c - b^2\mathcal{L} :: \frac{b}{3} + \frac{bc^2}{5a^2} + \frac{bc^4}{7a^4}, \&c. : \frac{a}{3} + \frac{ac^2}{15a^2} + \frac{ac^4}{35a^4}, \&c.,$$

that is, as

$$\frac{d - x}{3} + \frac{4dx \times \overline{d - x}}{5 \times (d + x)^2} + \frac{16d^2x^2 \times \overline{d - x}}{7 \times (d + x)^4}, \&c.$$

to

---

[59]In the published essay [68] the second part of the expression for $V/\mathcal{A}$ is given as (in different notation) $(\mathcal{L} - c)adc^{-3}(2a^2 + b^2)^{-1}$ and in the second part of the expression for $V/\frac{1}{2}(\mathcal{A} + \mathcal{B})$ the 4 is replaced by 2. I believe that these expressions in [68] are wrong; clearly there could be typographical errors. See NPIV, pp. 178–179, for details of the calculations.

$$\frac{d+x}{3} + \frac{4dx \times \overline{d+x}}{15 \times (d+x)^2} + \frac{16d^2x^2 \times \overline{d+x}}{35 \times (d+x)^4}, \&c.,$$

and so (terms having been neglected where higher powers of x appear) as[60]

$$\frac{1}{3}d + \frac{17}{15}x : \frac{1}{3}d + \frac{19}{15}x.$$

Thus there will be

$$\mathcal{B} - \mathcal{A} : \mathcal{B} + \mathcal{A}(= 2G) :: x : 5d + 18x$$

and

$$\mathcal{B} - \mathcal{A} : G :: 2x : 5d + 18x.$$

But by Cor. 2 of Prop. I, x is to d as $\mathcal{B} - \mathcal{A} + 3V$ to $\mathcal{B} + \mathcal{A} - 2V$, and so, on substituting the values of the quantities $\mathcal{B} - \mathcal{A}$ and $\mathcal{B} + \mathcal{A}$, there will be

$$x : d :: \frac{2Gx}{5d + 18x} + 3V : 2G - 2V.$$

Hence

$$2Gx - 2Vx = \frac{2Gdx + 15Vd^2 + 54Vdx}{5d + 18x},$$

and

$$10Gdx - 10dVx + 36Gx^2 - 36Vx^2 = 2Gdx + 15Vd^2 + 54Vdx,$$

and when all terms in which x^2 is found have been omitted, there will be

$$8Gdx - 64dVx = 15Vd^2 \quad \text{and} \quad x : d :: 15V : 8G - 64V,$$

and $2x$ is to d as $15V$ to $4G - 32V$. Therefore the rise of all the water, i.e., the difference of the semidiameters CA and CB (or $2x$) is to the mean semidiameter as $15V$ to $8G$ approximately; moreover, it will be easy to express this ratio more accurately whenever the application will demand it, by taking more terms of the value of the logarithm $\mathcal{L}$ and following the calculation through; furthermore, in this way x to d comes out more accurately as $15V$ to $8G - 57\frac{5}{14} \times V$.

Cor. $\mathcal{B} - \mathcal{A}$ is equal to $\frac{3V}{4}$ and $\mathcal{B} - G = \frac{3V}{8}$ approximately. For

$$\mathcal{B} - \mathcal{A} : G :: 2x : 5d :: 30V : 40G \quad \text{and so} \quad \mathcal{B} - \mathcal{A} : V :: 3 : 4.$$

Schol. It will be shown in the same way that the gravity at the pole B towards an oblate spheroid will be to the gravity at any place A on the equator, as

$$2CB \times CA \times \overline{CF - CS} \quad \text{to} \quad CA^2 \times CS - CB^2 \times CF.$$

---

[60]MacLaurin's expression for $\mathcal{A}/\mathcal{B}$ appears to be correct, but I believe that the approximation which he now gives is incorrect and that the subsequent deductions from it have to be amended; the assertions of the Corollary remain valid (see NPIV, pp. 180–181).

(p.182)

PROPOSITION V.

PROBLEM.

To find the force V which arises from the unequal gravity of the parts of the Earth towards the Sun, and to determine the rise of the water resulting from this.

Let S be the Sun, T the Earth, $ABab$ the lunar orbit, its eccentricity having been neglected, and B and b the quadratures. Let S denote the periodic time of the Earth about the Sun, L the periodic time of the Moon about the Earth, l the time in which the Moon would revolve about the Earth in the circle at mean distance Td $(= \frac{1}{2}CA + \frac{1}{2}CB)$ if the motion of the Moon were not disturbed by its gravity towards the Sun and were kept in orbit only by its gravity towards the Earth. Further, let K denote the mean gravity of the Moon or the Earth towards the Sun, g the gravity of the Moon towards the Earth at its mean distance, v the force which the action of the Sun would add to this gravity in the quadratures at the same distance. These things having been put in place, there will be $v : K :: dT :$

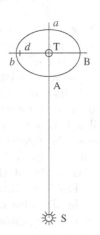

MacLaurin's Fig. 10

ST; and by the common doctrine of centripetal forces $K : g :: \frac{ST}{S^2} : \frac{dT}{l^2}$; hence $v : g :: l^2 : S^2$; and since l^2 is a little less than L^2, because the Moon is drawn away somewhat from the Earth by its gravity into the Sun, it is clear that the force v is to g in a ratio which is a little smaller than L^2 to S^2. Now, so far no one (as far as I have found out) has accurately determined this ratio of the force v to g; however, it seems to be more like the ratio L^2 to $S^2 + 2L^2$, or perhaps the ratio L^2 to $S^2 + \frac{3}{2}L^2$, than the ratio L^2 to S^2. In fact, mindful of the warning of the most illustrious Academy, I consider it appropriate to omit here the arguments by which it is inferred, since in this investigation it is of little importance which of these ratios is used. Let us suppose therefore with Newton that[61] $v : g :: L^2 : S^2 ::$ (by the astronomical calculations of the periods of the Sun and Moon) $1 : 178,725$. The force V which on the surface of the Earth corresponds to the force v, is to v, as the mean semidiameter of the Earth to the mean distance of the Moon, or as 1 to $60\frac{1}{2}$. Moreover, the force g acts along lines which are concurrent at the centre of gravity of the Earth and the Moon; when the ratio of this has been obtained from the increment of gravity in the descent to the surface of the Earth, it will be seen that the force V is to G (by which the mean gravity at the surface of the Earth is denoted as above) as 1 to 38604600. Hence,

---

[61]The comma which appears in the final quantity 178,725 (and in some subsequent expressions) represents in a common notation of the time a decimal point.

since by Cor. 2 of Prop. I there is $x : d :: 15V : 8G - 57\frac{5}{14}V$, there will be in this case $x : d :: 1 : 20589116$. And since the mean semidiameter of the Earth is 19615800 feet, it follows from this that the total rise of the water resulting from the force of the Sun will be one Parisian foot along with $\frac{90545}{100000}$ parts of a foot, that is to say, one foot along with ten inches and $\frac{8654}{10000}$ parts of an inch; Newton found this briefly by his method to be one foot eleven inches along with $\frac{1}{30}$th part of an inch, and this height differs from ours by only the sixth part of one inch.

Now in this calculation the Earth is supposed to be spherical, except in so far as the sea is raised up by the force of the Sun. But if we look for the greatest rise of the water, it has to be assumed that the Sun moves about the equator, and I have constructed the figure $ABab$ in this plane, and the force V has to be increased in the ratio of the mean semidiameter to the greatest semidiameter of the Earth, and the force G has to be decreased until it becomes equal to the gravity at the equator: that is to say, if we suppose the figure of the Earth to be that which Newton found, the force V will have to be increased in the ratio 459 to 460, and G will have to be reduced in almost the same ratio, since the forces of gravity on the surface of the Earth are inversely as the distance of the places from the centre; and since the distance d has to be increased in the same ratio, the rise of the water at the equator will have to be increased in the cube of the ratio of the mean semidiameter to the maximum, and so it will be one foot eleven inches along with about a 60th part of an inch. Now the Earth is higher at the equator than comes out by Newton's calculation under the hypothesis that the Earth is uniformly dense from the surface all the way to the centre, as is to be inferred from various observations of pendulums and especially from the measurement of a meridional degree which the most distinguished men have recently determined with the greatest accuracy at the polar circle.

SCHOL. 1. If we had taken the gravity to be equal at A and B, and of the same force in the whole circumference $ABab$, then x would have turned out to be equal to $\frac{3Vd}{2G}$ only, and the rise of the water (or $2x$) to one foot six inches along with about a third of an inch. Of course under this hypothesis CA to CB would have turned out to be as $G + V$ to $G - 2V$, and so x to d would have been as $\frac{3V}{2}$ to G approximately. And hence the use of the preceding Propositions is apparent, since the rise of the water according to the latter less accurate hypothesis is less that the rise which we have determined in this Proposition, the difference being $\frac{3Vd}{4G}$, namely, a quarter of the former rise.

SCHOL. 2. From this doctrine it is clear that the satellites of Jupiter must cause huge motions in the Ocean of Jupiter (if there is any) when they are in conjunction with or opposed to the Sun and each other, provided they are not much smaller than our Moon; for the diameter of Jupiter to the distance of any of its satellites has a much greater ratio than the diameter of the Earth to the distance of the Moon. It is probable that the mutations of the spots of

Jupiter, which have been observed by the astronomers, affect the rise at least in some respect; for if these mutations are discovered to preserve that analogy with the aspects of the satellites, which this doctrine requires, it will be proof that their true cause is to be sought from this source. From this doctrine we may also conjecture not without profit that the motions of satellites about their axes and about the primaries are so made up that they always show the same hemisphere to their primaries, according to the opinion of the most celebrated astronomers. For it is probable that very great motions of the sea must be caused in the satellites, if they were to revolve about their axes with any other velocity; but the tides arising from the different distances of the satellites from their primaries can suffice for setting in motion the waters in these (if there are any).

SECTION IV.

Concerning the motion of the Sea, in so far as it is altered by the daily motion of the Earth or by other causes.

(p.184)   We showed in the preceding Section that a fluid Earth unequally heavy towards the Sun or the Moon must take on the figure of an oblong spheroid; its transverse axis would pass through the centre of the luminary, if the Earth did not rotate about its axis with a daily motion; and we have determined the rise of the water resulting from the force of the Sun under the hypothesis that the Earth is at rest. But because of the motion of the Earth the system of the tide of the sea is different. For consequently, the water may never be in equilibrium, but is stirred up by uninterrupted motions. Let us suppose the Sun and the Moon to be in conjunction or opposed in the equatorial plane $ABab$; let Aa be the diameter which passes through their centres and Bb the perpendicular diameter to this (Fig. 1). While the mass of water is revolved by the daily motion, the forces by which its rise is brought about in the transit of the water from the places b and B to A and a are increased, and turn out to be greatest in the latter places; however, the rise of the water is seen to be continued after these forces have begun to decrease almost up to the places where these forces are equal to the forces by which it is depressed below the height which it would attain naturally, if the motion of the water were disturbed by no external force; thus the motion of the water can be considered as more even, and it rises almost as much under the forces by which it is raised while they are decreasing, as during their increase. And since the centrifugal force produced by the daily motion is much less than the gravity, the position of the place F where the previously mentioned forces are equal at the equator, while the water crosses over from the place b to the place A, seems able to be determined approximately as follows. Let Ff be the perpendicular from the point F to Bb and let fz be perpendicular to TF. Let V denote the sum of the forces with which the Sun and the Moon depress the water in the lines TB, Tb as above, and the force with which the water is raised up at F will be

$$\frac{3V \times Fz}{d} = \frac{3V \times Ff^2}{d \times TF}.$$

Let us suppose F to be the position of the water where the height of the water is least, so that TF can be taken for the conjugate semiaxis of the figure $ABab$, the gravity at the extremity of this axis being called $\mathcal{B}$ and the mean gravity in this figure G, as above; and the force with which the water is depressed below its natural position at the place F will be

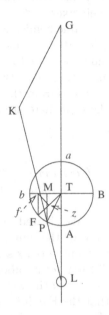

MacLaurin's Fig. 1

$$\mathcal{B} - G + \frac{V \times TF}{d}.$$

Let these forces be put equal, and since TF is approximately equal to the distance d, and $\mathcal{B} - G = \frac{3V}{8}$ by the Cor. of Prop. IV, there will be

$$\frac{3V}{8} + V = \frac{3V \times Ff^2}{d^2},$$

or

$$TF^2 : Ff^2 :: 3 : 1 + \tfrac{3}{8} :: 24 : 11,$$

hence the angle FTb will be 42 degrees 37 minutes, and it will occur almost at the middle point between b and A. But we do not put forth this calculation as accurate.

(p.186)

PROPOSITION VI.

PROBLEM.

To compare with each other the motion of the sea arising from the force of the Sun, and the lunar motion in an orbit which is approximately circular, and hence to estimate the rise of the water.

It is very well known to the astronomers that the mean distance of the Moon in the syzygies is less than the mean distance in the quadratures. The most illustrious Halley concluded from observations that the former distance is to the latter as $44\frac{1}{2}$ to $45\frac{1}{2}$. By a certain method of his own Newton found the ratio of these to be 69 to 70: Prop. 28 of Book 3 of the *Principia*. The most illustrious author of the Treatise on the Motions of the Moon according to the Theory of Gravity, who was very well informed in this doctrine, deduced it to be the number 69 to 70 without taking account of the decrease in gravity while the Moon crosses from the syzygies to the quadratures. In order that the motion of the sea arising from the force of the Sun (its nature is determined above in Prop. V) may be compared with the motion of the Moon, let us suppose the lunar globe to be filled up with water and let us investigate the rise of this water by means of Propositions IV and V. In Prop. V there was force v to g as 1 to 178,725; consequently in this case there would be

$$x : d :: 15v : 8g - 57\tfrac{5}{14} \times v :: 1 : 91,496 :$$

and so the semiaxis of the figure to the conjugate semiaxis (or $d + x$ to $d - x$) would be as 46,248 to 45,248, which almost coincides with the ratio of the distances of the Moon in the quadratures and the syzygies which Halley determined from observations; thus the figure of the lunar orbit is of a type scarcely different from that which the quiescent aqueous globe of the Moon would take on while completing its orbit under the force of the Sun; however, they would be in different positions, since the minor axis of the former looks back at the Sun, while the major axis of the latter is directed towards the Sun. The ratio of the numbers 59 to 60 (whose difference[62] is to the semisum as $3v$ to g approximately) agrees well with the ratio of the semiaxes of the figure which the water would take on as a result of the force of the Sun, if the force of gravity were the same over the whole circumference $ABab$, as we have shown in Schol. 1 of Prop. V. Moreover, the rise of the water determined in Prop. V agrees with that which Halley determined from observations; hence it may be conjectured that the difference of the diameters of the lunar orbit is made a little greater as a result of the decrease of the gravity of the Moon into the Earth while it crosses from the syzygies to the quadratures, in almost the same ratio as the rise of water has come out larger in this Proposition on account of the excess of the gravity of the water into the Earth at the place B over its gravity at the place A and at other distances from the centre. But whatever is to be decided about the ratio of the diameters of the lunar orbit, it may be deduced from these things that the rise of the water determined in Prop. V comes out scarcely larger on account of the daily motion of the Earth about its axis. For let us suppose this motion to be increased until the centrifugal force arising from this motion becomes equal to the gravity, and the particles of the sea are revolved in the manner of satellites in orbits which are almost circular and tangential to the Earth. These orbits will be ellipses whose minor axes produced will pass through the Sun. And if the difference of the semiaxes is to the mean semidiameter as $3V$ to G (according to those things which the most acute man teaches concerning the lunar motions), it will be less than the rise of the water determined above in Prop. V, in which we found $2x$ to be to d as $15V$ to $4G$. For if we investigate the difference of these semiaxes from the figure of the lunar orbit, since it is known from observations by the most distinguished Halley, it will exceed the rise of the water determined above just a little. And it is not to be wondered at if they do not agree accurately, since the gravity of the Moon towards the Earth

---

[62]The original has *semidifferentia*. See NPVI, footnote (83), p. 187.

follows the inverse ratio squared of the distances, while the gravity of the water is also greater at the smaller distance, but not in the same ratio. Since these phenomena are analogous and shed some light on each other, it seemed worthwhile to mention these things concerning their comparison with each other. However, we suppose here that the motion of the water continues in the same circle parallel to the equator, or the latitude is kept the same in individual revolutions, and we do not consider the variation of the rise of the water which results from the spheroidal figure of the Earth.

(p.187)

PROPOSITION VII.

The motion of the water is disturbed by the unequal velocity with which bodies are carried about the axis of the Earth by the daily motion.

Indeed, if a mass of water is carried by the tide, or some other cause, to a greater or lesser distance from the equator, it will fall into water which is carried with a different velocity about the axis of the Earth; hence the motion of the former is necessarily disturbed. The difference of the velocities with which bodies which are, for example, at a place located at 50° from the equator and at a place located only 36 miles further north, is greater than that which would be represented by 7 miles in individual hours, as will be clear from an easy calculation. And since the motion of the sea is sometimes so great that the tide will mark off 6 miles or even more in individual hours, the effects that can arise from this are not to be regarded as insignificant.

If the water is carried away from the south towards the north by the general motion of the tide, or by some other cause, the flow of the water will consequently deflect a little towards the east, since the water was first carried by the daily motion towards this region with a greater velocity than that which corresponds to a place located more towards the north. On the other hand, if the water is carried down from the north towards the south, the flow of the water will be deflected towards the west for a similar reason. And we suspect that various phenomena of the motion of the sea arise from this. For example, this may be why the icebergs which come away from the Arctic Ocean are more often observed in the western than in the eastern region of the Atlantic Ocean. But it is probable that greater tides can also be set in motion in many more places than those which result from a calculation of the forces of the Sun and Moon when the latitude has been taken into account. We suspect that the same cause serves for the generation of winds, especially the more violent ones, and sometimes for increasing or decreasing them, and for producing other phenomena both of the atmosphere and of the sea. But it is not permitted to pursue these things individually at the present time.

PROPOSITION VIII

PROBLEM.

To find the change in the rise of the water determined in Prop. V,
which results from the spheroidal figure of the Earth.

Let $PApa$, $PBpb$ be sections of the Earth through the poles P and p, the
first of which passes through the places A and a, where the height of the water
at the equator under the forces of the Sun and
Moon is greatest, while the latter passes through
the places B and b where it is least; let these
sections be ellipses, with F a focus of the figure
$PApa$, f a focus of the section $PBpb$, and g a
focus of the section $ABab$. And if all the sections of the solid which pass through the line Aa
are supposed to be elliptical we will find by a
calculation set up according to Lemma V that
the gravity at the place A towards this solid will
be to the gravity at the same place towards the
sphere with centre C described on the diameter
Aa as

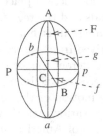

MacLaurin's Fig. 11

$$1 + \frac{3CF^2 + 3Cg^2}{10CA^2} + \frac{9CF^4 + 6CF^2 \times Cg^2 + 9Cg^4}{56CA^4}, \&c. \quad \text{to} \quad \frac{CA^2}{CB \times CP};$$

and if the gravity at the place B is determined by a similar calculation, by
use of the same Lemma and the Schol. of Prop. II, the ratio of the gravity
at A to the gravity at B will be established, and by Cor. 2 of Prop. I the
difference of the semidiameters CA and CB, or the rise of the water, will be
determined. However we omit the calculation, which is rather extended, since
it is of little use. I only wished to show by this Proposition that Geometry
would not fail us in a very celebrated problem which requires to be examined
with the greatest accuracy. But there remains a particular difficulty in this
investigation, about which a few things require to be added.

PROPOSITION IX.

To find the force of the Moon for moving the sea.

It is not possible to determine these things from the celestial motions, but
if the rise of the water in the syzygies of the luminaries, which is generated
by the sum of the forces of the Sun and Moon, is compared with the rise of
the water in the quadratures, which results from their difference, the force of
the Moon will be found from the force of the Sun which is given by Prop. V.
Newton investigates this from observations undertaken by Samuel Sturmy

before the mouth of the River Avon, from which he concludes that the rise of the water in the equinoctial syzygies is to the rise of the water in the same quadratures, as 9 to 5. Then after various calculations he concludes that the force of the Moon is to the force of the Sun as 4.4815 to 1, and that the rise of the water resulting from both forces in the mean distances of the luminaries will be 50 feet along with a half. We have investigated the ratio of these forces from observations reported by the celebrated Cassini in the work cited above. But since in addition to the general causes already mentioned, some of which can scarcely be reduced to calculation, various others depending on the location of places, the natural properties of the sea beds, the force of the winds and the region make the tides of the sea sometimes greater, sometimes smaller, it is not to be wondered at if the forces of the Moon which are derived from observations set up at different places, or at the same place but at different times, do not entirely agree. Therefore we will not spend time at present in listing the calculations which we have undertaken concerning the motion of the sea resulting from the force of the Moon. But after some observations which we are awaiting concerning the tides of the sea on the shores of America and the East Indies have come to hand, we may perhaps make a more informed judgement about these matters. We observe only that the tide seems to decrease in a ratio smaller than the square of the sine of the complement of the declination; also the remaining general laws of the tide are thrown into confusion by the forward and backward motion of the water. But we are afraid that weariness may result if we repeat what has already been treated by others long ago. Anomalous tides seem to depend for the most part on the location of places and seas. However, it is to be observed that it follows from the theory of gravity that only one tide must occur sometimes in the space of 24 hours in places beyond 62 degrees of latitude, if reciprocation of the motion of the water should permit it. *

Therefore if the analysis of the various causes which come together to produce the phenomena of the tide could be set up accurately, it would certainly contribute not a little to a more fruitful knowledge of the forces and motions of the system of the World. For the location of the centre of gravity of the Moon and the Earth and those things which concern the precession of the equinoxes and other distinguished phenomena of nature, would thus become known with greater certainty. For these reasons we have considered the accurate determination and demonstration of the amount of the rise of the water, as far as it may be understood from the celestial motions, under the assumption of the laws of gravity which are deduced from observations (this is not the place for an examination of its cause). We now willingly submit

---

* For let the declination of the Moon be 28 degrees and that of the place beyond 62 degrees towards the same region, and it is clear that the Moon comes into contact with the horizontal of this place only once in the space of 24 hours.

these thoughts of whatever kind to the judgement of the most illustrious ROYAL ACADEMY, which we always revere with all honour and respect.

REMARKS ON THE DISSERTATION
Concerning the Physical Cause of the Flow and Ebb of the Sea,
to which is prefixed the motto,
Opinionum commenta delet dies, Naturae judicia confirmat.

(p.189)   I. In Prop. IV it was found that $x = \frac{15Vd}{8G}$ approximately, which is a sufficiently accurate value of x, and it does not require any correction especially in the calculation of Prop. V. But it is more accurate that x is to d as $15V$ to $8G - \frac{88}{7}V$ and not as $15V$ to $8G - \frac{803}{14}V$ or $8G - 57\frac{5}{14}V$ as I had written at the end of Prop. IV by some error of the pen or calculation, which is certainly of little importance, and does not change the arguments of the subsequent Propositions. However, let me add here the main steps of the calculation. I had found in Prop. IV that B is to A as

$$\frac{1}{3} + \frac{c^2}{15a^2} + \frac{c^4}{35a^4}, \&c. \quad \text{to} \quad \frac{b}{a} \times \left(\frac{1}{3} + \frac{c^2}{5a^2} + \frac{c^4}{7a^4}, \&c. \right),$$

and so (on substituting in place of $\frac{b}{a}$ its value $\frac{\sqrt{a^2-c^2}}{a}$, or $1 - \frac{c^2}{2a^2} - \frac{c^4}{8a^4}, \&c.$) as

$$\frac{1}{3} + \frac{c^2}{15a^2} + \frac{c^4}{35a^4}, \&c. \quad \text{to} \quad \frac{1}{3} + \frac{c^2}{30a^2} + \frac{c^4}{840a^4}, \&c.,$$

from which it follows that $B - A$ is to G (or $\frac{1}{2}B + \frac{1}{2}A$) as

$$\frac{c^2}{10a^2} + \frac{23c^4}{8 \times 35a^4}, \&c. \quad \text{to} \quad 1 + \frac{3c^2}{20a^2} + \frac{25c^4}{8 \times 70a^4}, \&c..$$

Moreover, $c^2 = 4dx$, and $a^2 = d^2 + 2dx + x^2$ from those things which are supposed in the Proposition; hence

$$\frac{c^2}{4a^2} = \frac{x}{d} - \frac{2x^2}{d^2} + \frac{3x^3}{d^3}, \&c.$$

and on substituting in place of $\frac{c^2}{a^2}$ its value $\frac{4x}{d} - \frac{8x^2}{d^2}, \&c.$ it will come out that $B - A$ is to G approximately as $14dx + 18x^2$ to $35d^2 + 21dx - 17x^2$. And since by the Cor. of Prop. I

$$\overline{B - A} \times d + 3Vd = 2Gx - 2Vx - \frac{3Vx^2}{d},$$

let the value of $B - A$ be substituted and let the terms in which Vx^2 appears be neglected (since V is very small relative to G) and there will be

$$3 \times 35Vd^2 = 56Gdx - 133Vdx + 24Gx^2 \quad \text{and} \quad x = \frac{3 \times 35Vd^2}{56dG - 133Vd + 24Gx},$$

so that if for x in the denominator is written the value $\frac{15Vd}{8G}$, which is close to the true value, the more accurate value

$$\frac{3 \times 35Vd}{56G - 88V}$$

will result, and there will be $x : d :: 15V : 8G - \frac{88}{7}V$ approximately. By a slightly different method there comes out

$$x = \frac{15Vd}{8G} + \frac{165V^2d}{56G^2}, \&c.,$$

a series which it is not difficult to continue, whenever it seems worth the effort. In Prop. VI we investigated the figure of the water filling the lunar orb which arises from the action of the Sun. When this correction has been used and other things have been kept as before, the minor axis of the figure will be to the major as 46.742 to 47.742, which differs little from the ratio which we showed in that Proposition.

(p.191)   II. The series which we showed in Prop. VIII is deduced by Lemma V and Prop. II.

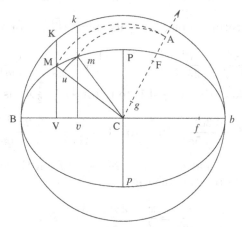

Figure from end of "Remarks"

Let $CA = a$, $CB = b$, $CP = e$, $CF = c$, $Cf = f$, $Cg = g$. Let ACM, ACm be any sections of the solid which pass through the line AC (which is normal to the plane $BPbp$). Let the arc mu drawn with centre C and radius Cm meet the line CM in u, and let the ordinates MV, mv meet the axis Bb in V and v, and the circle BKb in K and k. Let $CA^2 - CM^2 = x^2$, or x is the distance of the focus from the centre in the figure ACM, let $\mathcal{L}$ be the logarithm of the quantity $a\sqrt{\frac{a+x}{a-x}}$, and the ultimate ratio of the gravity of the particle A into the part bounded by the planes ACM, ACm to the gravity into the part of the sphere drawn with centre C and radius CA which is bounded by the same planes, will be as $3CM^2 \times \overline{\mathcal{L} - x}$ to x^3 by Prop. II. Therefore the gravity of the particle A into the solid will be as

$$\int \frac{3CM^2 \times \overline{\mathcal{L} - x}}{x^3} \times \frac{mu}{CM} = \int \frac{3CM \times mu}{x^3} \times \overline{\mathcal{L} - x}$$

$$= \int \frac{3CK \times Kk \times CP}{CK \times x^3} \times \overline{\mathcal{L} - x} = \int \frac{3e \times Kk}{x^3} \times \overline{\mathcal{L} - x}.$$

Let $CV = u$. And there will be

$$u^2 + (b^2 - u^2) \times \frac{e^2}{b^2} - CM^2 = a^2 - x^2.$$

Hence

$$e^2 + \frac{b^2 - e^2}{b^2} u^2 = a^2 - x^2, \quad \text{and} \quad u^2 = \overline{a^2 - e^2 - x^2} \times \frac{b^2}{b^2 - e^2} = \overline{c^2 - x^2} \times \frac{b^2}{f^2}.$$

And so

$$KV^2 = b^2 - u^2 = b^2 - \frac{b^2}{f^2} \times \overline{c^2 - x^2} = b^2 \times \frac{\overline{f^2 + x^2 - c^2}}{f^2} - \frac{b^2}{f^2} \times \overline{x^2 - g^2}.$$

Moreover, $Kk : Vv :: CK : KV$. Thus

$$Kk = \frac{b\,du}{KV} = \frac{b^2}{f} \times \frac{-x\,dx}{\sqrt{c^2 - x^2} \times \frac{b}{f}\sqrt{x^2 - g^2}} = \frac{-bx\,dx}{\sqrt{c^2 - x^2} \times \sqrt{x^2 - g^2}}.$$

Consequently, the gravity of the particle A towards the solid will be as

$$\int \frac{-3ebx\,dx}{x^3\sqrt{c^2 - x^2} \times \sqrt{x^2 - g^2}} \times \overline{\mathcal{L} - x}.$$

But

$$\mathcal{L} - x = \frac{x^3}{3a^2} + \frac{x^5}{5a^4}, \&c.$$

Consequently, that gravity will be as

$$\int \frac{-3ebx\,dx}{3a^2\sqrt{c^2 - x^2} \times \sqrt{x^2 - g^2}} + \int \frac{-3ebx^3\,dx}{5a^4\sqrt{c^2 - x^2} \times \sqrt{x^2 - g^2}}, \&c.$$

Let $z^2 = x^2 - g^2$, and the first integral will be

$$\int \frac{-eb\,dz}{a^2\sqrt{c^2 - g^2 - z^2}},$$

and the second will be

$$\int \frac{-3ebx^2\,dz}{5a^4\sqrt{c^2-g^2-z^2}} = \int \frac{-3eb\,dz \times \overline{z^2+g^2}}{5a^4\sqrt{c^2-g^2-z^2}}.$$

Along with the subsequent integrals,[63] these are easily reduced to circular arcs. And hence the ratio of the gravity of the particle A towards this solid to the gravity towards the sphere constructed on the semidiameter CA will be as assigned in the Proposition, the terms of the series decreasing very rapidly provided CF, Cf and CG are sufficiently small. If g vanishes, this series will give the gravity towards the spheroid at the equator; but this is investigated more elegantly in Prop. III.

(p.195)   III. In Prop. IX we observed following Newton that the force of the Moon for moving the sea could be compared with the force of the Sun, by comparing the tides in the syzygies and the quadratures; the same ratio could be obtained by comparing the tides which take place in the syzygies of the luminaries at different distances of the Moon from the Earth, if the tides were exactly proportional to the forces by which they are produced. Let L denote the mean force of the Moon, S the mean force of the Sun, X and x two different distances of the Moon from the Earth in the equinoctial syzygies, Z and z the distances of the Sun from the Earth in the same syzygies, d and D the mean distances of both; and if the declination of the Moon is zero, and the tides are as the forces of the luminaries, or as

$$\frac{Ld^3}{X^3} + \frac{SD^3}{Z^3} \quad \text{and} \quad \frac{Ld^3}{x^3} + \frac{SD^3}{z^3},$$

hence, by comparison of the tides the ratio L to S would be revealed. For let the rise of the water in the former case be to the rise in the latter as m to n, and L will be to S as

$$\frac{mD^3}{z^3} - \frac{nD^3}{Z^3} \quad \text{to} \quad \frac{nd^3}{X^3} - \frac{md^3}{x^3}.$$

---

[63]The Latin here is *cum subsequentibus summis.*

Notes on Part III

Note on Section I (pp. 97–99). Maclaurin quotes various tidal measurements reported by Jacques Cassini and cites for these the *Mémoires de Mathematique et de Physique* for the years 1710, 1712 and 1713. Six articles by Cassini on this topic appear in these volumes [17–22]; however, all the quoted data are to be found in the last article. I have reproduced and translated the relevant parts of [22] in Appendix III.4 (pp. 202–205) in order to clarify what is being measured and when. The heights are all given in the old French units (see my Introduction, p. 92) and the distance of the Moon from the Earth is expressed as d *parts*, where

$$\frac{\text{Distance}}{\text{Mean Distance}} = \frac{d}{1{,}000} \, .$$

Note on Section II (pp. 99–103). Newton discusses the motion of planets, in particular the motion of the Moon and of the Earth, and the tidal effects due to the Moon and the Sun on the Earth's waters in Book III of the *Principia*, "The System of the World" [85] ([15, 63]). MacLaurin refers specifically to Newton's argument concerning the attraction of the Moon towards the Earth, with which Propositions III, IV, XXVI–XXVIII, in particular, are concerned. He also mentions Newton's determination of the Sun's tidal effects, for which the relevant reference is Proposition XXXVI (see also items (49–51) on pp. 590–592 of [15]).[64] Proposition XXIV contains a general discussion of the tides and mentions the high tide taking place at the second or third lunar hour, to which MacLaurin refers (see also item (38) on p. 581 of [15]).

MacLaurin makes brief mention of the French expedition to Lapland, which has already been noted with references in my Introduction, p. 90. In a footnote he comments on the possible variation in the obliquity of the Ecliptic, the Sun's path across the celestial sphere (see [55]); it does not seem relevant to pursue this here except to note that about 1740 MacLaurin addressed the Edinburgh Philosophical Society on this topic and his paper

---

[64]The supplementary material included at the end of Book III in [15] and likewise headed *The System of the World* is apparently an early version of Book III.

was subsequently published by the Society [71]. MacLaurin suggested that "its variations will principally depend on the position of Jupiter and Saturn to the sun and earth," the effect being greatest when Jupiter and Saturn are aligned with and on the same side of the sun and so cause the greatest displacement of the sun relative to the fixed centre of gravity of the planetary system.

Just before he states some approximations from Newton, which are discussed separately in the next note, MacLaurin mentions the quadrature of the circle and of the hyperbola in connection respectively with the figure of the Earth and the rise of the water. In Propositions II and III, which relate to the tides, he has to evaluate

$$\int \frac{1}{a^2 - z^2} \, dz = \frac{1}{2a} \left(\int \frac{1}{a - z} \, dz + \int \frac{1}{a + z} \, dz \right);$$

the corresponding curves are the hyperbolas $y(a-z) = 1$ and $y(a+z) = 1$. In the Scholia to these Propositions he states corresponding results which apply to the figure of Earth; the corresponding integrals can be evaluated in terms of the length of a circular arc (see NPII, NPIII).

Note on "a few things ... from Newton" (pp. 102–103). The source may be Propositions XXV and XXVI in Book III of Newton's *Principia*, where a related diagram appears and the approximation of TG by $3PM$ is stated (in different notation) and applied (see pp. 428–432 in [85] ([63]) or pp. 440–444 in [15]; also Articles 472, 686 of [69]). The Earth is to be regarded as a fluid sphere which is then elongated along one axis Aa as a result of the attraction of the Moon (or the Sun); the resulting figure is a spheroid generated by rotating a certain ellipse about the elongated axis.

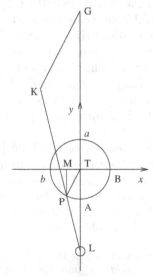

Under the inverse square law of attraction, if $\overrightarrow{TL}$ is chosen to represent the attraction of the Moon on a particle at the centre of the Earth and $\frac{|LK|}{|LT|} = \frac{|LT|^2}{|LP|^2}$, then $\overrightarrow{KL}$ will represent the attraction of the Moon at P. We then have $\overrightarrow{KL} = \overrightarrow{KG} + \overrightarrow{GL} = \overrightarrow{KG} + \overrightarrow{GT} + \overrightarrow{TL}$, where KG is parallel to PT; if P is at A or a then G has to be defined by continuity (see below). MacLaurin states the approximations $|KG| \approx |PT|$ and $|TG| \approx 3|PM|$. We note for the application of Proposition I that $\overrightarrow{KG}$ represents a force which acts at P towards the centre of the spheroid with magnitude approximately proportional to the distance of P from the centre, while $\overrightarrow{GT}$ represents a force which acts at P parallel to the axis, but away from the centre, with magnitude approximately proportional to three times the distance of P from the equatorial plane. The force represented by $\overrightarrow{TL}$ is ignored, since it is the same everywhere and so affects all particles of the spheroid equally. The same analysis will also apply when P lies inside the spheroid.

MacLaurin's diagram suggests that the generating figure is a circle, but it has to be an ellipse for the intended applications (see the Corollaries of Proposition I). We can derive the approximations analytically as follows. Introduce a coordinate system in MacLaurin's diagram as shown, let P be the point (x, y) on the lower semiellipse with equation $y = -a\sqrt{1 - \frac{x^2}{b^2}}$, where any one of $a > b$, $a = b$, $a < b$ $(a, b > 0)$ may hold, and let L, the centre of the Moon, be the point $(0, -l)$. Then

$$LP^2 = x^2 + \left(-a\sqrt{1 - \frac{x^2}{b^2}} + l\right)^2 = kx^2 + a^2 + l^2 - 2al\sqrt{1 - \frac{x^2}{b^2}},$$

where $k = (b^2 - a^2)/b^2$. Now we have $\frac{|LK|}{|LT|} = \frac{|LT|^2}{|LP|^2}$ by construction, and so

$$\frac{|LK|}{|LP|} = \frac{|LT|^3}{|LP|^3} = \frac{l^3}{\left(kx^2 + a^2 + l^2 - 2al\sqrt{1 - \frac{x^2}{b^2}}\right)^{3/2}},$$

and $\frac{|LP|}{|PK|} = \frac{|LT|}{|TG|}$, since PT and KG are parallel, therefore

$$|TG| = \frac{|LT| \cdot |PK|}{|LP|} = \frac{|LT| \cdot \big||LK| - |LP|\big|}{|LP|} = |LT|\left|\frac{|LK|}{|LP|} - 1\right|$$

$$= l\left| l^3 \left(kx^2 + a^2 + l^2 - 2al\sqrt{1 - \frac{x^2}{b^2}}\right)^{-3/2} - 1\right|$$

$$= l\left| \frac{l^3}{(a^2 + l^2)^{3/2}} \left(1 - \frac{2al\sqrt{1 - \frac{x^2}{b^2}} - kx^2}{a^2 + l^2}\right)^{-3/2} - 1\right|$$

$$\approx l \left| \frac{l^3}{(a^2+l^2)^{3/2}} \left(1 + \frac{3(2al\sqrt{1-\frac{x^2}{b^2}} - kx^2)}{2(a^2+l^2)} \right) - 1 \right|$$

$$= \left| \frac{3l^5 a}{(a^2+l^2)^{5/2}} \sqrt{1 - \frac{x^2}{b^2}} + l \left(\frac{l^3}{(a^2+l^2)^{3/2}} - 1 - \frac{3l^3 kx^2}{2(a^2+l^2)^{5/2}} \right) \right|$$

$$= \left| \frac{3l^5 |PM|}{(a^2+l^2)^{5/2}} + l \left(\frac{l^3}{(a^2+l^2)^{3/2}} - 1 - \frac{3l^3 kx^2}{2(a^2+l^2)^{5/2}} \right) \right| .$$

It is clear that

$$\frac{3l^5}{(a^2+l^2)^{5/2}} \to 3 \quad \text{as} \quad l \to \infty$$

and it is not difficult to see that

$$l \left(\frac{l^3}{(a^2+l^2)^{3/2}} - 1 - \frac{3l^3 kx^2}{2(a^2+l^2)^{5/2}} \right) \to 0 \quad \text{as} \quad l \to \infty .$$

Thus, on the assumption that $l \gg a$ and $l \gg b$, we may take

$$|TG| \approx 3|PM| .$$

We also have

$$\frac{|KG|}{|PT|} = \frac{|LG|}{|LT|} = \frac{|LT|+|TG|}{|LT|} = 1 + \frac{|TG|}{|LT|} \approx 1 + \frac{3|PM|}{|LT|} \approx 1$$

provided $l \gg a$ and $l \gg b$, which justifies MacLaurin's assertion that KG "is almost equal to PT itself."

Similar arguments produce similar results when P is on the upper semi-ellipse $y = a\sqrt{1 - \frac{x^2}{b^2}}$. Note that there will be a point on the arc APb shown where $|LP| = |LT|$ and therefore $|LK| = |LT|$; as P moves round arc Aba we will have $|LK| > |LP|$ up to this point and $|LK| < |LP|$ after it.

For the applications of these approximations we need to consider the special cases where P coincides with A, a, B or b.

Case (i): P at A ($x = 0$).

The point K now lies on LT (produced) and $\overrightarrow{KG}$, $\overrightarrow{GT}$ have opposite directions. Moreover,

$$|TG| = l \left(l^3 (a^2+l^2-2al)^{-3/2} - 1 \right) = l \left(l^3(l-a)^{-3} - 1 \right) \approx 3a ,$$

so that

$$|LG| = |LT| + |TG| = l^4(l-a)^{-3} ,$$

and

$$|LK| = \frac{|LT|^3}{|LP|^2} = \frac{l^3}{(l-a)^2} .$$

Combining these, we obtain

$$|KG| = |LG| - |LK| = \frac{l^3 a}{(l-a)^3} \approx a \quad \text{and} \quad |TK| = |TG| - |KG| \approx 2a\,,$$

provided $l \gg a$.

Case (ii): P at B or b $(x = \pm b)$.
The line KG is now perpendicular to LG, and we have

$$|LK| = \frac{l^3}{kb^2 + a^2 + l^2} = \frac{l^3}{b^2 + l^2}\,,$$

$$|TG| = l\left(l^3(kb^2 + a^2 + l^2)^{-3/2} - 1\right) = l\left(l^3(b^2 + l^2)^{-3/2} - 1\right) \approx -\frac{3b^2}{2l}\,,$$

$$|LG| = |LT| + |TG| = l^4(b^2 + l^2)^{-3/2}\,,$$

$$|KG|^2 = |LK|^2 - |LG|^2 = \frac{b^2 l^6}{(b^2 + l^2)^3}\,.$$

Consequently,

$$|KG| = \frac{bl^3}{(b^2 + l^2)^{3/2}} \approx b \quad \text{and} \quad |TG| \approx 0\,,$$

provided $l \gg b$.

Finally:

Case (iii): P at a $(x = 0)$.
Now we are on the upper semiellipse and

$$|TG| = l\left(1 - l^3(a^2 + l^2 + 2al)^{-3/2}\right) = l\left(1 - l^3(l + a)^{-3}\right) \approx 3a\,,$$

with G lying below the line Bb. The forces represented by $\overrightarrow{GT}$ at A and a therefore have approximately the same magnitudes but opposite directions. Again $|KG| \approx a$, but the direction of $\overrightarrow{KG}$ is opposite to that at A, and $|TK| \approx 2a$.

Note on Lemma I (pp. 103–105). In the proof of Lemma I MacLaurin uses two properties of the ellipse, for which proofs can be found in his *Treatise of Fluxions*. They are first established for the circle and then deduced for the ellipse by projection (see Appendix III.1, pp. 197–199). Article 615 of [69] contains the following result (Fig. 1).

Let GH, KL be parallel chords in an ellipse and let a chord VR meet them in M and N, respectively. Then

$$\frac{GM \cdot MH}{KN \cdot NL} = \frac{VM \cdot MR}{VN \cdot NR}\,. \tag{1}$$

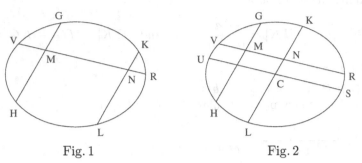

Fig. 1   Fig. 2

If we now let KL be the diameter parallel to GH and introduce the diameter US parallel to VR (Fig. 2), we can deduce the first property required in the proof of Lemma I as follows. Apply the above result to the parallel chords VR, US and the chord LK to get

$$\frac{VN \cdot NR}{UC \cdot CS} = \frac{KN \cdot NL}{KC \cdot CL} . \tag{2}$$

Then from (1) and (2) we deduce

$$\frac{GM \cdot MH}{VM \cdot MR} = \frac{KN \cdot NL}{VN \cdot NR} = \frac{KC \cdot CL}{UC \cdot CS} = \frac{CL^2}{CS^2} . \tag{3}$$

In words, the ratio of the rectangles formed by intersecting chords is equal to the ratio of the squares of the semidiameters parallel to the respective chords.

The second property used is found in Article 612 of [69].

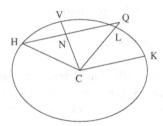

Let CK, CV and CH, CL be two pairs of conjugate semidiameters in an ellipse. Let the line through H parallel to CK meet CL in Q and CV in N. Then

$$NH \cdot HQ = CK^2 . \tag{4}$$

The proof of Lemma I also makes use of the similarity of triangles qmz and ECG and of triangles Hzu and gGC. Note that, although the Lemma is stated and proved for an arbitrary pair of conjugate diameters, the diagram (below) shows these as the axes of the ellipse, which is the situation required in Corollaries 3 and 4.

The last line in Corollary 1 perhaps requires some amplification. We have $mu = uM$ since mM is an ordinate to diameter HI. The parallel lines mx, *er*, QM must therefore cut off equal segments from PQ; thus $Qe = eq$ as MacLaurin asserts. From $\frac{Pz}{qz} = \frac{PQ}{qe}$ we obtain

$$\frac{Pq}{qz} = \frac{Pz - qz}{qz} = \frac{PQ - qe}{qe} = \frac{Pq + eQ}{qe} = \frac{Pq + qe}{qe} = \frac{Pe}{qe}.$$

We now use the following property of ratios: if $\frac{a}{b} = \frac{c}{d}$ then (where meaningful) $\frac{a+c}{b+d} = \frac{a-c}{b-d} = \frac{a}{b} = \frac{c}{d}$. Thus,

$$\frac{Pq}{qz} = \frac{Pe}{qe} = \frac{Pe - Pq}{qe - qz} = \frac{qe}{ze} = \frac{Pe + qe}{qe + ze} = \frac{Pe + eQ}{eQ + ze} = \frac{PQ}{zQ},$$

as required.

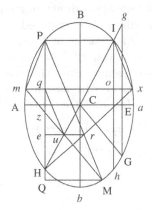

Diagram for Lemma I
MacLaurin's Fig. 2

In the proof of Corollary 2 we get $ox = qm$ because the semidiameter CB bisects the chord mx, which is parallel to the conjugate semidiameter CA, and also qo, since HP, CB, hI are parallel and C is the midpoint of HI. For the ratios note that $Io = Pq$ for similar reasons and by Corollary 1 and the similarity of triangles qzm and QzM

$$\frac{Pq}{qm} = \frac{PQ}{Qz} \cdot \frac{qz}{qm} = \frac{PQ}{Qz} \cdot \frac{Qz}{QM} = \frac{PQ}{QM}.$$

The triangles Iox and PQM are therefore similar, the angles at o and Q being equal, and then, since $Io \parallel PQ$ and $ox \parallel QM$, we must have $Ix \parallel PM$.

MacLaurin asserts next that the diameters parallel to Ix (and therefore also to PM) and Hx are conjugate. This is another general property of the ellipse which follows immediately from the corresponding property in the circle on projection (see diagrams below). In the case of the circle the lines corresponding to Ix and Hx must be at right angles and so the corresponding diameters are also at right angles; on projection these become conjugate diameters of the ellipse and parallel lines remain parallel (see Appendix III.1, pp. 197–199). The Corollary follows on applying the Lemma with HP, mx replaced by Hx, MP, the lines qu, PM becoming ru, xm, respectively.

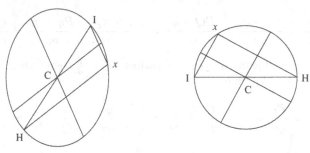

Corollary 3 assumes the situation shown in its diagram below, namely, Bb and Aa are the axes of the ellipse. It is required to show that $VN = Vn = er$ and $DN \parallel PM$, $Dn \parallel Pm$. MacLaurin first uses the similarity of triangles qmz and QMz along with Corollary 1 to show that triangles PQM and Pqm are similar and consequently that the angles QPM, or HPM, and qPm, or HPm, are equal. Using the similarity of triangles Pqm and Per and of triangles Her and Hqx and then property (3) discussed above, he obtains

$$\frac{He \cdot Pe}{er^2} = \frac{CB^2}{CA^2}. \tag{5}$$

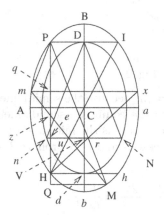

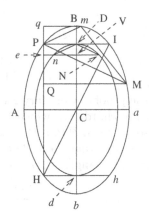

Diagram for Lemma I Cor. 3   Diagram for Lemma I Cor. 4
    MacLaurin's Fig. 3   Second Part

The second ellipse has centre C and Dd as one axis, the other axis being along Aa and having length $Aa \times \frac{Dd}{Bb}$. Since nN is parallel to the second axis we have $nV = VN$. Applying (3) in this ellipse and using the proportionality of the axes of the two ellipses, we get

$$\frac{DV \cdot Vd}{VN^2} = \frac{DV \cdot Vd}{nV \cdot VN} = \frac{CB^2}{CA^2}. \tag{6}$$

Clearly, $dV = He$ and $DV = Pe$, so we can deduce from (5) and (6) that $VN = er$. The triangles Per and DVN are therefore congruent, from which it follows that Pr and DN must be parallel. Finally, from

$$\angle mPH = \angle HPM = \angle VDN = \angle VDn\,,$$

we obtain $Pm \parallel Dn$.

It is clear from the diagram for Corollary 3 that

$$PQ + Pq = Pe + eQ + Pe - qe = 2Pe = 2DV\,, \tag{7}$$

since $eQ = qe$ (see above). This is the first part of Corollary 4, where Q and q lie on the same side of P, which is the situation while m lies between H and P. When m coincides with P, the line Dn will be parallel to the tangent to the ellipse at P; the point q will then coincide with P and we will have $PQ = 2DV$. As m continues round the ellipse between P and I we will have the configuration shown in the last diagram above. Equation (7) has to be changed to

$$PQ - qP = Pe + eQ - (qe - Pe) = 2Pe = 2DV\,, \tag{8}$$

which is the second part of Corollary 4.

Corollary 4 is crucial for the proof of Lemma IV, which in turn provides the basis for MacLaurin's Fundamental Theorem. It is perhaps surprising therefore that MacLaurin left the reader to verify it as a converse. The editors of [86] apparently thought so for they included a proof "by analysis" in their edition of MacLaurin's essay. I have given their proof in translation along with some explanatory notes in Appendix III.5, pp. 205–208.

Note on Lemma II (p. 106).

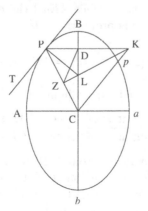

MacLaurin's Fig. 4

The circle on PL as diameter must pass through D and Z since $\angle PDL$ and $\angle PZL$ are right angles, and TP must be the tangent to this circle at P since $\angle TPL$ is a right angle. Then $\angle PDZ = \angle PLZ$ since they are angles subtended at the circumference of this circle by the same chord PZ and on the same side of it. Next $\angle PLZ = \angle TPZ$ because TP is the tangent at P to the circle, and $\angle TPZ = \angle PCK$ because $PT \parallel CK$. Now $\angle PDZ$ and $\angle ZDK$ add up to $180°$, so the same is true of $\angle ZDK$ and $\angle ZCK$; since these are opposite angles in the quadrilateral $CKDZ$, it must therefore be cyclic. Then, being angles subtended at the circumference of the circle through C, K, D, Z by the chord CK and on the same side of it, $\angle CZK = \angle CDK$, and $\angle CDK$ is a right angle by construction; it follows that KZ passes through L since both LZ and KZ are perpendicular to CP. Since PK, PC meet this circle in D, K and Z, C respectively, we have $PC \cdot PZ = PK \cdot PD$. Finally, $PK \cdot PD = CA^2$ by (4) in NLI, p. 142.

Note on Lemma III (pp. 106–107).

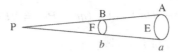

MacLaurin's Fig. 5

In the first instance, MacLaurin appears to be considering the intersection of a right circular cone with spheres centred at the vertex of the cone. In this case the area of the surface $AEaA$ is $2\pi|PA|^2(1 - \cos\alpha)$, where α is the angle between the axis of the cone and any generator. Then, under the inverse square law of attraction, the attraction of the surface $AEaA$ acts along the axis of the cone and for small α is approximately proportional to $2\pi(1 - \cos\alpha)$, which is independent of $|PA|$. The attraction at P of the whole conical figure bounded by the surface $AEaA$ therefore acts along the axis of the cone and its magnitude is proportional to

$$\int_0^{|PA|} 2\pi(1 - \cos\alpha)\,dr = 2\pi(1 - \cos\alpha)|PA|\,;$$

hence, with reference to MacLaurin's figure,

$$\frac{|\text{Attraction at } P \text{ of "cone" } PAEa|}{|\text{Attraction at } P \text{ of "cone" } PBFb|} = \frac{|PA|}{|PB|}\,,$$

as MacLaurin has it. Integrating from $|PB|$ to $|PA|$, we see more generally that the attraction at P of the portion R of the "cone" bounded by the surfaces $AEaA$ and $BFbB$ acts along the axis of the cone and its magnitude is proportional to $2\pi(1 - \cos\alpha)|BA|$. In fact the dependence on α is a little different: this attraction is given by (see equation (5) of Appendix III.3 (p. 202))

$$k \iiint_R \frac{z}{(x^2 + y^2 + z^2)^{3/2}} \, dx \, dy \, dz \, ,$$

where P is the origin and the z-axis is along the axis of the cone; transforming to spherical polar coordinates ($x = \rho \sin\phi \cos\theta$, $y = \rho \sin\phi \sin\theta$, $z = \rho \cos\phi$) or using equation (6) of Appendix III.3 gives

$$k \int_0^\alpha \int_0^{2\pi} \int_{|PB|}^{|PA|} \sin\phi \cos\phi \, d\rho \, d\theta \, d\phi = k\pi(|PA| - |PB|) \int_0^\alpha \sin 2\phi \, d\phi$$

$$= \frac{k\pi}{2} |BA| (1 - \cos 2\alpha) \, .$$

Clearly, the above arguments can be adapted to the situation where cones are replaced by right pyramids. As MacLaurin suggests in Corollary 1, he applies this Lemma by approximating a given solid by finite unions of such pyramidal figures, or portions of them (if, for example, the point P is not on the surface), with certain parameters tending to zero to realise the solid as the limit of these unions (see Lemmas IV–VI). Of course, the details are rather sketchy as the pyramids are in general neither regular nor bounded by spherical surfaces. The attraction at a point P from such a union will be the vector sum of the attractions from the individual pyramids or parts of them and the attraction at P of the solid will be the limit of these sums.

In the diagram below, $\mathcal{S}_1$, $\mathcal{S}_2$ represent similar homogeneous solids and the points P_1, P_2 and Q_1, Q_2 are similarly situated with respect to $\mathcal{S}_1$, $\mathcal{S}_2$, respectively; that is to say (in three dimensions), the second configuration can be obtained from the first by scaling, translation and rotation.

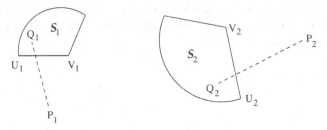

For each approximating solid for $\mathcal{S}_1$ there will be a similar, similarly situated approximating solid for $\mathcal{S}_2$ and by the Lemma the ratio of the magnitudes of the attractions at P_2, P_1 from the respective approximating solids will be equal to the scaling factor; this will also hold in the limit so that

$$\frac{|\text{Attraction at } P_2 \text{ from } \mathcal{S}_2|}{|\text{Attraction at } P_1 \text{ from } \mathcal{S}_1|} = \frac{|P_2 Q_2|}{|P_1 Q_1|} \, . \tag{1}$$

The scaling factor can be determined from any corresponding lengths in $\mathcal{S}_1$ and $\mathcal{S}_2$, so if, for example, $U_1 V_1$, $U_2 V_2$ are corresponding edges in $\mathcal{S}_1$, $\mathcal{S}_2$, respectively, the last ratio will also be equal to $|U_2V_2|/|U_1V_1|$. This is the content of Corollary 1.

In Corollary 2 MacLaurin first asserts that, if a hollow solid is formed by rotating the region bounded between two similar, similarly situated ellipses[65] about one of their axes, there is zero attraction from the solid at any point P on the inner surface or in the hollow interior. This he derives from the fact that, if RS is a chord in an ellipse and its intercepts with an interior, similar, similarly situated ellipse are T, U as shown below, then $RT = US$; referring to the diagram below, we can obtain this last property easily by noting that, if KX is the semidiameter in the outer ellipse bisecting RS, then by similarity KY must be the semidiameter in the interior ellipse bisecting TU.

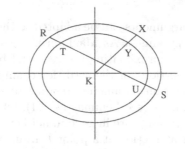

MacLaurin appears to consider only the generating ellipses, but in fact we have to take into account all sections of the solid by planes through P which are parallel to the axis of rotation or, equivalently, are perpendicular to the equatorial plane (see first figure below). Now, as MacLaurin notes in Article 633 of [69], "the sections of two similar concentric spheroids similarly situated, which are made by the same plane, are similar ellipses" (see Appendix III.2, pp. 199–201). Thus the sections we have to consider always consist of the region bounded by two similar, similarly situated ellipses as represented in the second figure below.

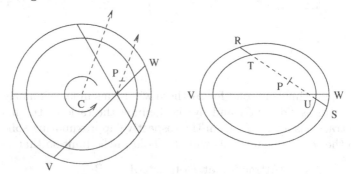

Effectively, two integrations produce the required result: we have as a limiting case of the Lemma that the segments RT and US produce attractions at P

---

[65]Such elipses have the same centre, corresponding axes lie along the same line, and the ratio of major axis to minor axis is the same in both cases.

which have opposite directions and equal magnitudes and therefore cancel each other out; consequently, considering all such chords RS through P, we deduce that the whole section produces zero attraction at P; then, taking all such sections into consideration, we see that the spheroidal shell itself has zero attraction at P.

We can easily make this precise by applying formula (3) of Appendix III.3 (p. 202) with the spheroidal shell as the region V and $P(x_0, y_0, z_0)$ any point inside or on the inner surface, the z-axis being the axis of rotation. In this case ϕ varies over $[0, \pi]$, while θ varies independently over $[0, 2\pi]$. As noted above, the spheroidal shell cuts off segments of equal length from a line through P ($RT = US$); these segments correspond to the pairs (ϕ, θ) and $(\pi - \phi, \theta + \pi)$. Thus by formula (4) of Appendix III.3 the integrals with respect to ρ in (3) for the pairs (ϕ, θ) and $(\pi - \phi, \theta + \pi)$ cancel out. Pairing the integrals with respect to ρ in this way as θ and ϕ vary independently over $[0, \pi]$, we see that the triple integral in (3) must have value zero.

Suppose now that B, D are points on the same semidiameter of an ellipse and take ellipses through B and D which are similar and similarly situated to the given ellipse.

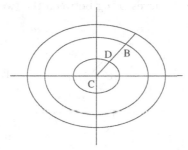

Let $\mathcal{E}$, $\mathcal{E}_B$, $\mathcal{E}_D$ be the solid spheroids obtained by rotating these three ellipses about the same axis. Since B, D are similarly situated on their corresponding ellipses, we have from Corollary 1 that

$$\frac{|\text{Attraction at } B \text{ from } \mathcal{E}_B|}{|\text{Attraction at } D \text{ from } \mathcal{E}_D|} = \frac{|CB|}{|CD|};$$

moreover, the first part of Corollary 2 shows that the attraction at B or D from the outer spheroid $\mathcal{E}$ is equal to the attraction at the point from $\mathcal{E}_B$ or $\mathcal{E}_D$, respectively, since in both cases the remaining portion of $\mathcal{E}$ produces zero attraction at the point. Thus

$$\frac{|\text{Attraction at } B \text{ from } \mathcal{E}|}{|\text{Attraction at } D \text{ from } \mathcal{E}|} = \frac{|CB|}{|CD|}.$$

As noted by Todhunter, much of the content of Lemma III and its Corollaries is implicit in Newton's *Principia* ([103], Articles 242–243, [85] ([15, 63]) Book I, Prop. XCI, Cor. III).

Note on Lemma IV (pp. 108–111). Note first of all that the axis of rotation Aa may be either the major or the minor axis of the generating ellipse. In the statement of the Lemma, MacLaurin resolves the force of attraction at P from the spheroid into two components, one parallel to the axis of rotation, the other perpendicular to it; of course, by the symmetry of the spheroid, the component perpendicular to these two directions must be zero. Corollary 4 of Lemma I is applied in the argument, and for this we need to note that, if two spheroids are generated by rotating similar, similarly situated ellipses about one of their axes, any plane which meets both of them cuts off similar, similarly situated ellipses from them. This is shown by MacLaurin in Article 633 of [69] (see Appendix III.2, pp. 199–201).

There is some notational confusion. In the statement of the Lemma, ABa denotes the generating ellipse, but in the demonstration it becomes an arbitrary section of the spheroid by a plane containing the line PDI – this will in fact be similar to the generating ellipse ([69], Article 633, see Appendix III.2). The point C is the centre of this varying ellipse and not the centre of the spheroid itself, which will be denoted by O below. In an attempt to clarify MacLaurin's diagram, lines and curves are drawn unbroken in one plane and dashed in the other; curves going between the two planes are dotted (see diagram opposite).

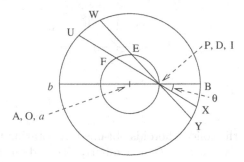

(Looking along axis of rotation)

Fig. 1

The spheroid is split up by planes through PDI (Fig. 1), which are perpendicular to the plane generated by the rotation of the axis Bb, the equatorial plane. MacLaurin considers the attraction at P of the wedges UWP and XYP between successive planes. He compares this with the attraction at D of the corresponding wedge FED of the interior spheroid. We take the angle between successive planes to be $\frac{\pi}{2k}$.

The inner ellipse (see diagram opposite) is split up by a succession of lines DN, DN' and Dn, Dn', where $\angle dDN = \angle dDn$ and $\angle nDn' = \angle NDN' = \frac{\pi}{2k}$, and k pairs of pyramids are formed with vertices at D and bases $NN'R'R$, $nn'r'r$ by taking planes through DN, DN', Dn, Dn' which are perpendicular to the plane of the ellipse. By Lemma III the attraction at D of such a pair of pyramids will be approximately proportional to $\overrightarrow{DN}$ and $\overrightarrow{Dn}$, respectively, provided k is large; clearly $\overrightarrow{DN} + \overrightarrow{Dn} = 2\overrightarrow{DV}$. The union of the k pairs of pyramids will be the inner wedge.

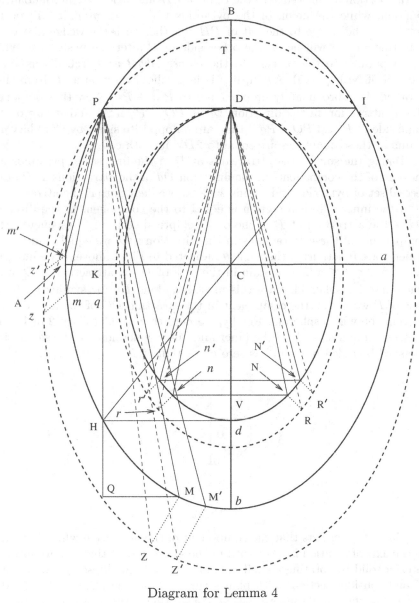

Diagram for Lemma 4
MacLaurin's Fig. 6

Corresponding pairs of pyramids are formed with vertices at P and bases $MM'Z'Z$, $mm'z'z$ by taking PM, PM', Pm, Pm' parallel to DN, DN', Dn, Dn', respectively. The union of the pyramids with bases $MM'Z'Z$ will be the portion of the wedge to the right of PH and below PI in MacLaurin's diagram, while the union of the pyramids with bases $mm'z'z$ will be the portion of the wedge to the left of PH together with the wedge above PI: note that m, m' will lie on the other side of P after the position in which Dn is parallel to the tangent to the ellipse $AbaB$ at P (cf. discussion at the end of NLI, p. 145). Again, if k is large the attraction at P from these pyramids is approximately proportional to $\overrightarrow{PM} + \overrightarrow{Pm}$. Now the component of this attraction in the direction $\overrightarrow{Dd}$ is $PQ + Pq$ if M and m are on the same side of P and $PQ - Pq$ if they are on opposite sides; by Corollary 4 of Lemma I these quantities are equal to $2DV$ in both cases.

Taking the sum of the attractions at D of the first set of pyramids and the sum of the components in the direction $\overrightarrow{Dd}$ of the attractions at P of the second set of pyramids and letting $k \to \infty$, we deduce that the attraction at D of the inner elliptical section is equal to the component in the direction $\overrightarrow{Dd}$ of the attraction at P of the outer elliptical section. Consequently, the components of these attractions in the direction $\overrightarrow{DO}$ are equal. These components are functions of the angle θ indicated in Fig. 1 above. If we integrate over $-\pi/2 \le \theta \le \pi/2$ we obtain in the case of D the attraction at D of the whole interior spheroid, since this must act along $\overrightarrow{DO}$ by symmetry. In the case of P we obtain the component in the direction $\overrightarrow{DO}$ of the attraction at P from the whole spheroid. Finally, we note that by Corollary 2 of Lemma III the attractions at D of the inner and outer spheroids are the same, since the shell bounded by them has zero attraction at D.

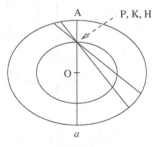

Fig. 2

MacLaurin asserts that the component of the attraction which is parallel to the axis of rotation is dealt with in the same way. In this case we form the interior solid by rotating a similar, similarly situated ellipse passing through K and consider sections with planes through the line PKH (Fig. 2). After discussing the corollaries we will establish Maclaurin's results by applying formula (6) of Appendix III.3 (p. 202).

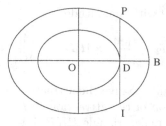

Fig. 3

It follows from Corollary 2 of Lemma III that *all* points on the line PI (Fig. 3) have the same component of attraction in the direction $\overrightarrow{DO}$. Consequently, by symmetry, the magnitude of the component of the attraction at a point in the spheroid along the perpendicular from the point to the axis is the same for all points at the same distance from the axis. Moreover, this magnitude is always proportional to the perpendicular distance: this follows from the above and the final part of Corollary 2 of Lemma III, which implies that

$$\frac{|\text{Attraction at } D \text{ towards inner spheroid}|}{|\text{Attraction at } B \text{ towards outer spheroid}|} = \frac{|DO|}{|BO|},$$

and therefore

$$|\text{Attraction at } D \text{ towards inner spheroid}| = (\text{constant}) \times |DO|.$$

This is part of Corollary 1 of Lemma 4. The corresponding assertions concerning the component perpendicular to the equatorial plane are dealt with similarly.

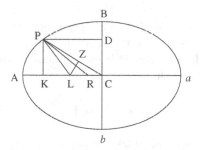

Diagram for Lemma IV Cor. 2
(from MacLaurin's Fig. 7)

In Corollary 2 of Lemma IV the term *middle circle* denotes the equator, the circle traced out by B as the ellipse is rotated about its axis Aa. From the discussion of Corollary 1 we have that the attraction at P is made up of two components:

(i) one along $\overrightarrow{PK}$ with magnitude

$$m_1 = \frac{|\mathcal{B}|}{|BC|} \times |PK| = \frac{|\mathcal{B}|}{|BC|} \times |DC|;$$

(ii) the other along $\overrightarrow{PD}$ with magnitude

$$m_2 = \frac{|\mathcal{A}|}{|AC|} \times |PD| = \frac{|\mathcal{A}|}{|AC|} \times |KC|.$$

These combine vectorially to give the attraction at P. Thus the direction of the attraction at P will be along $\overrightarrow{PR}$ if we make triangle PKR similar to the adjacent triangle with PK, KR corresponding to m_1, m_2, respectively, that is to say,

$$\frac{|PK|}{|KR|} = \frac{m_1}{m_2} = \frac{|\mathcal{B}| \times |DC|}{|BC|} \times \frac{|AC|}{|\mathcal{A}| \times |KC|}, \quad \text{or} \quad \frac{|KR|}{|KC|} = \frac{|\mathcal{A}|}{|AC|} \times \frac{|BC|}{|\mathcal{B}|},$$

as MacLaurin has it. Moreover, by comparing the hypotenuses of the two triangles, we see that, if m is the magnitude of the attraction at P, then

$$\frac{m}{|PR|} = \frac{m_1}{|PK|} = \frac{m_2}{|KR|}.$$

Finally, we show how Lemma IV may be established using formula (6) of Appendix III.3 (p. 202); in fact we have precisely the limiting case of MacLaurin's geometric argument which was discussed above. We may take for the spheroid the equation

$$\frac{x^2}{\alpha^2} + \frac{y^2}{\alpha^2} + \frac{z^2}{\beta^2} = 1, \tag{1}$$

where $0 < \alpha \le \beta$ or $0 < \beta \le \alpha$. This represents the spheroid obtained by rotating the ellipse $y^2/\alpha^2 + z^2/\beta^2 = 1$ about the z-axis. By symmetry, or by rotation of the coordinate axes, it is clearly enough to establish the Lemma for a point $P(0, y_0, z_0)$ with

$$y_0 \ge 0, \quad z_0 \ge 0, \quad \text{and} \quad \frac{y_0^2}{\alpha^2} + \frac{z_0^2}{\beta^2} = 1. \tag{2}$$

The appropriate unit vectors which are perpendicular to the axis and parallel to it are $(0, -1, 0)$ and $(0, 0, -1)$, respectively, and so, from the cited formula, we have to consider

$$-\iiint_{V'} \sin^2\phi \sin\theta \, d\rho \, d\phi \, d\theta \quad \text{and} \quad -\iiint_{V'} \sin\phi \cos\phi \, d\rho \, d\phi \, d\theta;$$

here V' is the set of (ρ, ϕ, θ) values which describe the solid spheroid, where

$$x = \rho \sin\phi \cos\theta, \quad y = y_0 + \rho \sin\phi \sin\theta, \quad z = z_0 + \rho \cos\phi.$$

By substituting these in (1) and using (2) we obtain

$$\rho = \rho(\phi,\theta) = -\frac{2(\beta^2 y_0 \sin\phi \sin\theta + \alpha^2 z_0 \cos\phi)}{\beta^2 \sin^2\phi + \alpha^2 \cos^2\phi}. \tag{3}$$

Note that the whole spheroid will be generated if we let ρ vary from 0 to $\rho(\phi,\theta)$ while ϕ and θ vary independently over $[0,\pi]$ (this requires us to admit negative values of ρ).

For the component perpendicular to the axis we have, for some constant k,

$$-k \int_0^\pi \int_0^\pi \int_0^{\rho(\phi,\theta)} \sin^2\phi \sin\theta \, d\rho \, d\phi \, d\theta$$

$$= 2k \int_0^\pi \int_0^\pi \frac{\beta^2 y_0 \sin\phi \sin\theta + \alpha^2 z_0 \cos\phi}{\beta^2 \sin^2\phi + \alpha^2 \cos^2\phi} \sin^2\phi \sin\theta \, d\phi \, d\theta \qquad \text{(by (3))}$$

$$= 2k \int_0^\pi \int_0^\pi \frac{\beta^2 y_0 \sin^3\phi \sin^2\theta}{\beta^2 \sin^2\phi + \alpha^2 \cos^2\phi} \, d\phi \, d\theta$$

(the other part of the integrand is an odd function of ϕ about $\pi/2$ for each θ)

$$= 2k\beta^2 y_0 \left[\frac{\theta}{2} - \frac{1}{4}\sin 2\theta\right]_0^\pi \int_0^\pi \frac{\sin^3\phi}{\beta^2 \sin^2\phi + \alpha^2 \cos^2\phi} \, d\phi$$

$$= k\pi\beta^2 y_0 \int_0^\pi \frac{\sin^3\phi}{\beta^2 \sin^2\phi + \alpha^2 \cos^2\phi} \, d\phi. \tag{4}$$

When $y_0 = \alpha$, so that $z_0 = 0$, this becomes

$$k\pi\alpha\beta^2 \int_0^\pi \frac{\sin^3\phi}{\beta^2 \sin^2\phi + \alpha^2 \cos^2\phi} \, d\phi. \tag{5}$$

Now the similar, similarly situated spheroid through the point $D(0,y_0,0)$ has equation

$$\frac{x^2}{y_0^2} + \frac{y^2}{y_0^2} + \frac{\alpha^2 z^2}{\beta^2 y_0^2} = 1, \tag{6}$$

and so we deduce from (5), on replacing α with y_0 and β with $\beta y_0/\alpha$, that the corresponding component of the attraction at D from this spheroid is

$$k\pi y_0 \frac{\beta^2 y_0^2}{\alpha^2} \int_0^\pi \frac{\sin^3\phi}{\frac{\beta^2 y_0^2}{\alpha^2} \sin^2\phi + y_0^2 \cos^2\phi} \, d\phi$$

$$= k\pi\beta^2 y_0 \int_0^\pi \frac{\sin^3\phi}{\beta^2 \sin^2\phi + \alpha^2 \cos^2\phi} \, d\phi,$$

which is the same (4). By symmetry, the other components of attraction at D due to the inner spheroid (6) are zero.

For the component parallel to the axis we have

$$- k \int_0^\pi \int_0^\pi \int_0^{\rho(\phi,\theta)} \sin\phi \cos\phi \, d\rho \, d\phi \, d\theta$$

$$= 2k \int_0^\pi \int_0^\pi \frac{\beta^2 y_0 \sin\phi \sin\theta + \alpha^2 z_0 \cos\phi}{\beta^2 \sin^2\phi + \alpha^2 \cos^2\phi} \sin\phi \cos\phi \, d\phi \, d\theta \qquad \text{(by (3))}$$

$$= 2k\alpha^2 z_0 \int_0^\pi \int_0^\pi \frac{\sin\phi \cos^2\phi}{\beta^2 \sin^2\phi + \alpha^2 \cos^2\phi} \, d\phi \, d\theta$$

(the other part of the integrand is an odd function of ϕ about $\pi/2$ for each θ)

$$= 2k\pi\alpha^2 z_0 \int_0^\pi \frac{\sin\phi \cos^2\phi}{\beta^2 \sin^2\phi + \alpha^2 \cos^2\phi} \, d\phi . \qquad (7)$$

When $z_0 = \beta$, so that $y_0 = 0$, this becomes

$$2k\pi\alpha^2\beta \int_0^\pi \frac{\sin\phi \cos^2\phi}{\beta^2 \sin^2\phi + \alpha^2 \cos^2\phi} \, d\phi . \qquad (8)$$

Now the similar, similarly situated spheroid through the point $K(0,0,z_0)$ has equation

$$\frac{\beta^2 x^2}{\alpha^2 z_0^2} + \frac{\beta^2 y^2}{\alpha^2 z_0^2} + \frac{z^2}{z_0^2} = 1 ,$$

and, as above with D, we deduce from (8), on replacing α with $\alpha z_0/\beta$ and β with z_0, that the corresponding component of the attraction at K from this spheroid is the same as that given in (7); again the other components at K are necessarily zero.

In Propositions II and III and their associated Lemmas V and VI MacLaurin obtains versions of (5) and (8) and evaluates the integrals. To show the connection with this later work let us conclude this Note by observing that by means of the substitution $u = \beta \cos\phi$ we obtain

$$\int_0^\pi \frac{\sin\phi \cos^2\phi}{\beta^2 \sin^2\phi + \alpha^2 \cos^2\phi} \, d\phi = \frac{2}{\beta} \int_0^\beta \frac{u^2}{\beta^4 + (\alpha^2 - \beta^2)u^2} \, du , \qquad (9)$$

while $u = \alpha \cos\phi$ gives

$$\int_0^\pi \frac{\sin^3\phi}{\beta^2 \sin^2\phi + \alpha^2 \cos^2\phi} \, d\phi = \frac{2}{\alpha} \int_0^\alpha \frac{\alpha^2 - u^2}{\alpha^2\beta^2 + (\alpha^2 - \beta^2)u^2} \, du . \qquad (10)$$

The transformed integral in (9) appears in Proposition II (case $\alpha < \beta$) and its Scholium (case $\alpha > \beta$); that in (10) occurs in Proposition III (case $\alpha < \beta$) and its Scholium (case $\alpha > \beta$). (See NPII (pp. 170–171), NPIII (pp. 174–177).)

Note on Proposition I (pp. 111–115). As MacLaurin's heading, *Fundamental Theorem*, suggests, Proposition I is the basis of much of the subsequent material in the dissertation. The conditions on the external forces correspond to those identified earlier for the representing vectors $\overrightarrow{KG}$ and $\overrightarrow{GT}$ (see Note on "a few things ... from Newton," pp. 138–139). Note, however, that a force acting towards the centre with magnitude proportional to the distance from the centre can be resolved into two forces, one acting parallel to the axis with magnitude proportional to the distance from the equatorial plane, the other acting perpendicular to the axis with magnitude proportional to the distance from the axis. In the version in Articles 636–640 in [69] MacLaurin recasts the external forces as forces parallel or perpendicular to the axis with magnitudes proportional to the appropriate distances. Centrifugal force, which is obviously important in the study of the figure of the Earth, is an example for the perpendicular case. As Todhunter states ([103], Article 245) MacLaurin only shows the possibility of equilibrium under his hypotheses by verifying that certain *necessary* conditions are satisfied.

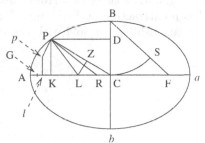

Diagram for Prop. I
MacLaurin's Fig. 7

The first condition to be satisfied is due to Huygens and is sometimes referred to as the *principle of the plumb-line* ([103], Article 53). Now, at any point P on the surface of the spheroid three forces are acting, the attraction of the spheroid at P and the two external forces. Each of these can be resolved into components perpendicular to the axis Aa and parallel to it. Each component perpendicular to the axis is proportional to $|PK|$ and each component parallel to the axis is proportional to $|PD|$: in the case of the attraction this comes from Corollary 1 of Lemma 4, for the first external force this has already been noted above, and for the second external force it is given (its component perpendicular to the axis is always zero). Thus the resultant force at P can be resolved into two components, F_1 perpendicular to the axis and proportional to $|PK|$ and F_2 parallel to the axis and proportional to $|PD|$. At A (or a) these components are 0 and M, respectively, and at B (or at any point on the equator) they are N and 0, respectively. Thus

$$\frac{F_1}{N} = \frac{|PK|}{|BC|} \quad \text{and} \quad \frac{F_2}{M} = \frac{|PD|}{|AC|}.$$

Moreover, by hypothesis, $M \times |CA| = N \times |CB|$. Thus, for P not at the equator, we have[66]

$$\frac{F_1}{F_2} = \frac{N \times |PK|}{|BC|} \times \frac{|AC|}{M \times |PD|} = \frac{|AC|^2}{|BC|^2} \times \frac{|PK|}{|KC|} = \frac{|KC|}{|KL|} \times \frac{|PK|}{|KC|} = \frac{|PK|}{|KL|}.$$

It follows that the resultant force at P acts along $\overrightarrow{PL}$, which is normal to the surface. Moreover, since $F_1 = k|PK|$, where the constant $k = N/|BC|$, we deduce that

$$F_2 = \frac{|KL|}{|PK|}k|PK| = k|KL|,$$

and the resultant force at P has magnitude

$$\sqrt{F_1^2 + F_2^2} = k\sqrt{|PK|^2 + |KL|^2} = k|PL|.$$

The above discussion can be applied at an interior point P if the spheroid is replaced by the similar, similarly situated spheroid through P (Fig. 1) – recall that the outer shell has no attraction at P (Lemma III, Cor. 2). Exactly the same proportionality relationships will hold with the same value of k.

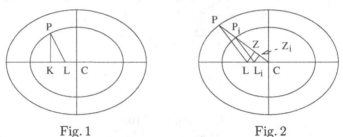

Fig. 1   Fig. 2

   The second condition is Newton's *principle of balancing columns* ([85] ([15, 63]), Book III, Propositions XIX, XX, [103], Article 23). Here, a column is a line from a point P on the surface of the spheroid to its centre C and the object is to show that all such columns have the same "weight." Consider an arbitrary point P_i on PC (Fig. 2). As noted above, the resultant force at P_i will be determined by the line P_iL_i which is normal to the similar, similarly situated spheroid on whose surface P_i lies; moreover, P_iL_i will be parallel to PL (the tangents at P and P_i must be parallel by similarity; alternatively, apply the methods of Appendix III.1, pp. 197–199). Thus, if LZ, L_iZ_i are the perpendiculars from L, L_i, respectively, to PC,

$$\frac{|P_iZ_i|}{|PZ|} = \frac{|P_iL_i|}{|PL|} = \frac{|P_iC|}{|PC|} \quad \text{and so} \quad |P_iZ_i| = \frac{|PZ|}{|PC|} \times |P_iC|.$$

---

[66]Here MacLaurin uses a general property of the ellipse, namely, that $\frac{|KC|}{|KL|} = \frac{|CA|^2}{|CB|^2}$ (see Diagram for Prop. I); this is easily established using the equations for the ellipse and a normal to it.

The component of the resultant force at P in the direction $\overrightarrow{PC}$ is therefore

$$k\frac{|PZ|}{|PC|} \times |P_iC|,$$

and the "weight" of the whole column is then the integral of this quantity with respect to $|P_iC|$ over the range 0 to $|PC|$, that is,

$$k\frac{|PZ|}{|PC|} \times \frac{1}{2}|PC|^2 = \frac{1}{2}k|PZ| \times |PC|.$$

By Lemma II, $|PZ| \times |PC| = |CB|^2$ and so the "weight" is the same for all columns from the surface to C.[67]
    The third condition is a generalisation of the second: MacLaurin claims that columns still balance if the centre C is replaced by any interior point of the spheroid. He begins by stating that the "weight" of a column from a point P on the surface to an interior point p is equal to (see MacLaurin's diagram for Prop. 1)

$$\frac{N}{2CB} \times \overline{PK^2 - pl^2} - \frac{M}{2CA} \times \overline{Cl^2 - CK^2}, \tag{1}$$

asserting that this may be established "by an easy calculation" which is omitted "for the sake of brevity." MacLaurin gives his proof of the third condition, including the derivation of the above expression, in Article 639 of [69], but he only provides details for the case where p and P are in the same meridian plane as in his diagram, claiming, "In like manner it is shown that any other columns from the surface of the spheroid to the particle p press equally upon it, and sustain each other." According to Todhunter it is not obvious how this could be done by MacLaurin's methods ([103], Article 245); it may therefore have been the case that MacLaurin did not have a convincing general proof. The notation and diagram of Article 639 are different from what we have here. I have adapted the argument to the present situation; the details are given below and are followed by a general proof.
    In Fig. 3 below (cf. Fig. 288 in [69]) P represents an arbitrary point on the surface of the spheroid, Pf is an arbitrary line through P which lies in a section of the spheroid by a plane containing the axis of rotation Aa, and p is initially a varying point which lies in the spheroid and on the line; further, the lines PD, pg are perpendicular to Bb and PK, pl are perpendicular to Aa, while ge, lu are perpendicular to Pf. Now the resultant force at p has components $M \times |pg|/|AC|$ in the direction $\overrightarrow{pg}$ and $N \times |pl|/|BC|$ in the direction $\overrightarrow{pl}$. On projecting onto Pf we see that the component of the resultant force at p in the direction $\overrightarrow{Pp}$ is

$$N \times \frac{|pu|}{|BC|} - M \times \frac{|pe|}{|AC|}.$$

---

[67]The axes have been interchanged between Lemma II and Proposition I, so we have $|CB|$ rather than the $|CA|$ of Lemma II.

Now, since triangles peg and pgh are similar, we have

$$\frac{|pe|}{|pg|} = \frac{|pg|}{|ph|}, \quad \text{so that} \quad |pe| = \frac{|pg|^2}{|ph|},$$

and, since triangles pul and pfl are similar, we obtain likewise

$$|pu| = \frac{|pl|^2}{|pf|}.$$

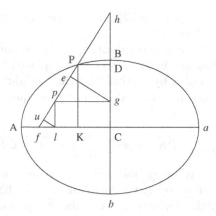

Fig. 3

The component of the resultant force at p in the direction $\overrightarrow{Pp}$ is therefore

$$\frac{N}{|BC|} \times \frac{|pl|^2}{|pf|^2} \times |pf| - \frac{M}{|AC|} \times \frac{|pg|^2}{|ph|^2} \times |ph|$$

$$= \frac{N}{|BC|} \times \frac{|pl|^2}{|pf|^2} \times (|Pf| - |Pp|) - \frac{M}{|AC|} \times \frac{|pg|^2}{|ph|^2} \times (|Pp| + |Ph|). \quad (2)$$

Note that $|pl|/|pf|$ and $|pg|/|ph|$ remain constant as p varies on the given line. To obtain the "weight" of the column from P to any point p which is in the spheroid and on the line we integrate (2) with respect to $|Pp|$ over the range 0 to $|Pp|$ to get

$$\frac{N}{2|BC|} \times \frac{|pl|^2}{|pf|^2} \times (|Pf|^2 - |pf|^2) - \frac{M}{2|AC|} \times \frac{|pg|^2}{|ph|^2} \times (|ph|^2 - |Ph|^2)$$

$$= \frac{N}{2|BC|} \times (|PK|^2 - |pl|^2) - \frac{M}{2|AC|} \times (|lC|^2 - |KC|^2),$$

since

$$\frac{|Pf|}{|pf|} = \frac{|PK|}{|pl|}, \quad |pg| = |lC|, \quad \frac{|Ph|}{|ph|} = \frac{|PD|}{|pg|}, \quad |PD| = |KC|.$$

This corresponds to MacLaurin's stated expression (1), which he simplifies by using the relations

$$N = M \times \frac{|CA|}{|CB|}, \quad \frac{|PK|^2}{|CA|^2 - |CK|^2} = \frac{|CB|^2}{|CA|^2}, \quad \frac{|pl|^2}{|CG|^2 - |Cl|^2} = \frac{|CB|^2}{|CA|^2}.$$

The first of these is a hypothesis, while the other two correspond to a variant of the canonical equation of the ellipse: if the generating ellipse has equation

$$\frac{x^2}{a^2} + \frac{y^2}{b^2} = 1, \quad \text{then} \quad \frac{y^2}{a^2 - x^2} = \frac{y^2}{a^2 \times \frac{y^2}{b^2}} = \frac{b^2}{a^2},$$

which produces the second identity; the similar, similarly situated ellipse through p has equation

$$\frac{x^2}{(\lambda a)^2} + \frac{y^2}{(\lambda b)^2} = 1,$$

for some λ, giving

$$|GC| = |\lambda a|, \quad \text{and} \quad \frac{y^2}{(\lambda a)^2 - x^2} = \frac{(\lambda b)^2}{(\lambda a)^2} = \frac{b^2}{a^2},$$

from which the third follows. The "weight" of the column Pp turns out to be

$$\frac{M(|CA|^2 - |CG|^2)}{2|CA|},$$

which is independent of P.[68]

To establish the result in general we may assume for the spheroid an equation of the form

$$\frac{x^2}{b^2} + \frac{y^2}{b^2} + \frac{z^2}{a^2} = 1;$$

here we are rotating the ellipse in the z, y-plane with equation $y^2/b^2 + z^2/a^2 = 1$ about the z-axis. Let $p(x_p, y_p, z_p)$ be an interior point of the spheroid and let $P(x_P, y_P, z_P)$ lie on its surface. The line segment joining P to p has equation

$$(x, y, z) = (x_P, y_P, z_P) + t(x_p - x_P, y_p - y_P, z_p - z_P) \quad (0 \le t \le 1).$$

Let T be the point on $\overrightarrow{Pp}$ with parameter t and let T', T'' be its projections on the axis of rotation and on the equatorial plane, respectively, so that

$$\overrightarrow{TT'} = -(x_P, y_P, 0) - t(x_p - x_P, y_p - y_P, 0), \quad \overrightarrow{TT''} = -(0, 0, z_P) - t(0, 0, z_p - z_P).$$

The resultant force at T is made up of the forces

---

[68]In the original the weight is given as $(M \times CA - M \times CG)/2$. This is either a typographical error or an error in simplification, and I have corrected it in the translation (see Part (3) of MacLaurin's proof). MacLaurin's previous, unsimplified expression is correct.

$$\frac{M}{a}\Big(-(0,0,z_P) - t(0,0,z_p - z_P)\Big),$$

which acts in the direction $\overrightarrow{TT''}$, and

$$\frac{N}{b}\Big(-(x_P, y_P, 0) - t(x_p - x_P, y_p - y_P, 0)\Big)$$

$$= \frac{aM}{b^2}\Big(-(x_P, y_P, 0) - t(x_p - x_P, y_p - y_P, 0)\Big),$$

which acts in the direction $\overrightarrow{TT'}$. As suggested by the reference to integrals above, the "weight" of the column Pp has to be interpreted as the work done in going from P to p along the line segment $\overrightarrow{Pp}$. This is

$$\int_0^1 \Big(\frac{aM}{b^2}\Big(-(x_P, y_P, 0) - t(x_p - x_P, y_p - y_P, 0)$$

$$+ \frac{M}{a}\Big(-(0,0,z_P) - t(0,0,z_p - z_P)\Big)\Big) \cdot (x_p - x_P, y_p - y_P, z_p - z_P)\, dt$$

$$= \int_0^1 \frac{aM}{b^2}\Big(-x_P(x_p - x_P) - y_P(y_p - y_P) - t\big((x_p - x_P)^2 + (y_p - y_P)^2\big)\Big)$$

$$+ \frac{M}{a}\Big(-z_P(z_p - z_P) - t(z_p - z_P)^2\Big)\, dt$$

$$= \frac{aM}{b^2}\Big(-x_P(x_p - x_P) - y_P(y_p - y_P) - \frac{1}{2}\big((x_p - x_P)^2 + (y_p - y_P)^2\big)\Big)$$

$$+ \frac{M}{a}\Big(-z_P(z_p - z_P) - \frac{1}{2}(z_p - z_P)^2\Big)$$

$$= \frac{aM}{2b^2}\big(x_P^2 + y_P^2 - x_p^2 - y_p^2\big) + \frac{M}{2a}\big(z_P^2 - z_p^2\big)$$

$$= \frac{aM}{2}\Big(\frac{x_P^2}{b^2} + \frac{y_P^2}{b^2} + \frac{z_P^2}{a^2} - \frac{x_p^2}{b^2} - \frac{y_p^2}{b^2} - \frac{z_p^2}{a^2}\Big)$$

$$= \frac{aM}{2}\Big(1 - \frac{x_p^2}{b^2} - \frac{y_p^2}{b^2} - \frac{z_p^2}{a^2}\Big),$$

which is independent of P. Finally, we note that, if $|CG|/|CA| = \lambda$, the similar, similarly situated spheroid through p has equation

$$\frac{x^2}{(\lambda b)^2} + \frac{y^2}{(\lambda b)^2} + \frac{z^2}{(\lambda a)^2} = 1,$$

so we may write the last expression as

$$\frac{M|CA|}{2}\left(1 - \frac{|CG|^2}{|CA|^2}\right) = \frac{M(|CA|^2 - |CG|^2)}{2|CA|},$$

as before.

In Corollary 1 MacLaurin considers the effect of the lunar (or solar) attraction on the fluid Earth. As noted above, the external forces are those forces represented by the vectors $\overrightarrow{KG}$ and $\overrightarrow{GT}$ discussed in the Note on "a few things ... from Newton."[69] Since KG is approximately proportional to the distance of P from the centre, if V denotes the magnitude of the corresponding force when this distance is $d = \frac{1}{2}(a + b)$, then at A this force has magnitude aV/d approximately and acts in the direction $\overrightarrow{AC}$ while at B its magnitude is bV/d approximately and it acts in the direction $\overrightarrow{BC}$. At A we have $|GT| \approx 3a \approx 3|KG|$, so $\overrightarrow{GT}$ represents a force acting in the direction $\overrightarrow{CA}$ with magnitude $3aV/d$ approximately. At B we have $|GT| \approx 0$. Thus, in MacLaurin's terminology and approximations,

$$M = \mathcal{A} + \frac{aV}{d} - \frac{3aV}{d} = \mathcal{A} - \frac{2aV}{d}, \tag{3}$$

$$N = \mathcal{B} + \frac{bV}{d}. \tag{4}$$

According to Proposition I, we require for equilibrium that $M/N = b/a$; hence

$$\frac{\mathcal{A} - \frac{2aV}{d}}{\mathcal{B} + \frac{bV}{d}} = \frac{b}{a},$$

and so

$$b\mathcal{B} + \frac{b^2V}{d} = a\mathcal{A} - \frac{2a^2V}{d}, \tag{5}$$

giving

$$a\mathcal{A} - b\mathcal{B} = \frac{2a^2V}{d} + \frac{b^2V}{d}, \tag{6}$$

as MacLaurin has it.

For Corollary 2 we put $a = d + x$, $b = d - x$ in (5) to obtain

$$(d - x)\mathcal{B} + \frac{(d-x)^2V}{d} = (d + x)\mathcal{A} - \frac{2(d+x)^2V}{d}. \tag{7}$$

On neglecting terms in x^2 this gives

$$d(\mathcal{B} - \mathcal{A} + 3V) \approx x(\mathcal{A} + \mathcal{B} - 2V), \quad \text{or} \quad \frac{x}{d} \approx \frac{\mathcal{B} - \mathcal{A} + 3V}{\mathcal{B} + \mathcal{A} - 2V}.$$

---

[69]See pp. 138–141. Note that C now denotes the centre, which was previously T.

The relevance of this to the tides is that $2x = a - b$ is the maximum possible difference between the heights of the water at any point due to the attraction of the Moon as it performs an orbit about the Earth.

Corollary 3 is just a variant of Corollaries 1 and 2 to deal with the situation where the magnitude of the central force is specified at a distance other than $\frac{1}{2}(a + b)$.

In Corollary 4 MacLaurin deals with the combined effects of the Moon and the Sun when they are aligned with the Earth. The separate effects of the Moon and the Sun combine to give approximately the same results as in Corollaries 1 and 2 (in particular, (6) remains valid) if V now denotes the magnitude of the combined central forces at distance $d = \frac{1}{2}(a + b)$.

Corollary 5 is concerned with the quadratures, where the Sun is along the axis Bb perpendicular to LT. Separately, the attraction of the Moon elongates the axis Aa, while the attraction of the Sun elongates the axis Bb. MacLaurin argues that the effect of the Moon dominates, so we will again have a spheroid with axis of rotation Aa. The single force of magnitude V is now replaced by two forces, namely, the central force at distance $d = \frac{1}{2}(a+b)$ due to the Moon, whose magnitude is denoted by l, and the corresponding force due to the Sun with magnitude s. Note that the roles of A and B are interchanged for the Moon and the Sun. Thus we now have

$$M = \mathcal{A} - \frac{2al}{d} + \frac{as}{d}, \tag{8}$$

$$N = \mathcal{B} + \frac{bl}{d} - \frac{2bs}{d}. \tag{9}$$

In (8) (resp. (9)) we have the attraction at A (resp. B), the effect of the Moon at A (cf. (3)) (resp. B (cf. (4))) and the effect of the Sun at A (cf. (4)) (resp. B (cf. (3))). According to Proposition I we now require for equilibrium

$$\frac{\mathcal{A} - \frac{2al}{d} + \frac{as}{d}}{\mathcal{B} + \frac{bl}{d} - \frac{2bs}{d}} = \frac{b}{a}. \tag{10}$$

Writing $a = d + x$, $b = d - x$ and ignoring terms in x^2 as in Corollary 2 leads to

$$\frac{x}{d} \approx \frac{\mathcal{B} - \mathcal{A} + 3(l - s)}{\mathcal{B} + \mathcal{A} - 2(l - s) - 4s} \approx \frac{\mathcal{B} - \mathcal{A} + 3V}{\mathcal{B} + \mathcal{A} - 2V},$$

provided $s \ll l$, where $V = l - s$.

For later application in Proposition IV let us note that from (10) we obtain

$$a\mathcal{A} - b\mathcal{B} = \frac{2a^2(l - s)}{d} + \frac{b^2(l - s)}{d} + \frac{(a^2 - b^2)s}{d} \approx \frac{2a^2 V}{d} + \frac{b^2 V}{d},$$

provided $\frac{(a^2 - b^2)s}{d}$ is small. Thus equation (6) holds approximately in the quadratures.

In the Scholium[70] MacLaurin states some consequences of Proposition I which relate to the figure of the Earth. Rotation of a spheroid about its axis with constant angular velocity produces at each point in it a centrifugal force which is proportional to the distance of the point from the axis. In the case of a fluid spheroid where the external forces are the corresponding centrifugal forces we require for equilibrium

$$\frac{\mathcal{B}}{\mathcal{A} - V} = \frac{a}{b},$$

where V is the magnitude of the centrifugal force at the equator. The spheroid must be rotating about the minor axis Bb of the generating ellipse and the centrifugal force at any point on the axis is zero. This is discussed in Article 641 of [69]. The assertion about the measure of a degree in the meridian is established in Article 657 by showing that the radius of curvature at a point on the generating ellipse is proportional to the cube of the length of the normal from the point to the major axis.

---

[70]In the translation I have corrected some errors in the labelling of points and lines (cf. [68]).

Note on Lemma V (pp. 116–117).

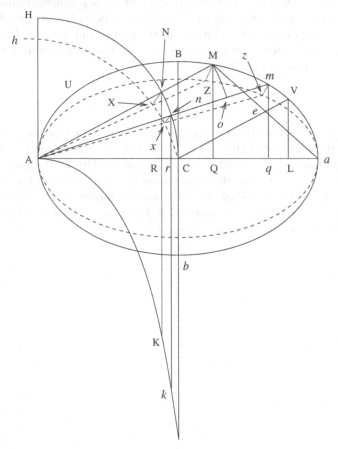

Diagram for Lemma V and Proposition II
MacLaurin's Fig. 8

As with the figure for Lemma IV, lines and curves in one plane are drawn
unbroken; those in the other are dashed, and curves going between the two
planes are shown as dotted. Although MacLaurin presents the Lemma as a
general result, the figure shows the intended application in Proposition II,
where $ABab$ is an ellipse which is to be rotated about its major axis Aa and
C is its centre. However, the graph (RK over AC) shown in MacLaurin's
original figure (see Appendix III.6, p. 210) is not correct for this case (cf. the
function obtained in Proposition II) and has been corrected

MacLaurin's argument that the component in the direction $\overrightarrow{AC}$ of the
attraction at A of the pyramid $ANXxn$ is proportional to $|Rr|$ seems to me
to require amplification. Certainly, by the ideas of Lemma III the magnitude
of the attraction at A of the pyramid is approximately equal to

$$k \times \frac{|NX| \times |Nn|}{|AN|^2} \times |AN| = k \times \frac{|NX| \times |Nn|}{|AN|}, \tag{1}$$

for an appropriate constant k.[71] Now the line segment $|XN|$ has length approximately equal to $|RN| \tan \alpha$, where α is the angle between the planes.[72] From calculus the length of an arc of the semicircle with equation $y = \sqrt{\rho^2 - x^2}$ is given by $\int \sqrt{1 + \frac{x^2}{\rho^2 - x^2}} \, dx = \rho \int \frac{1}{y} \, dx$ over the appropriate range of x values, so

$$\text{arc } Nn \approx |AN| \times \frac{1}{|RN|} \times |Rr|.$$

Substituting these expressions in (1) we obtain $k|Rr| \tan \alpha$ as an approximation to the magnitude of the attraction at A of the pyramid $ANXxn$; the magnitude of its component in the direction $\overrightarrow{AC}$ will then be approximately

$$\frac{k}{\rho} |AR||Rr| \tan \alpha \qquad \left(\cos \theta = \frac{1}{\rho} |AR| \right).$$

By Lemma III the component of the attraction at A of the pyramid $AMZzm$ in the direction $\overrightarrow{AC}$ therefore has approximate magnitude

$$\left(\frac{k}{\rho} |AR||Rr| \tan \alpha \right) \frac{|AQ|}{|AR|} = \frac{k}{\rho} |AQ||Rr| \tan \alpha = \frac{k}{\rho} |RK||Rr| \tan \alpha. \tag{2}$$

The circular arc Mo plays no explicit role in MacLaurin's proof after its introduction. Presumably, he intended to use it to approximate the pyramid $AMZzm$, which has spheroidal face, by one with spherical face which he can then compare directly with the pyramid $ANXxn$ (cf. MacLaurin's *Remarks* II and its Note, pp. 134–136, 191–195).

Now consider the spheroid to be sliced up by a succession of $4t$ half-planes bounded by the axis of rotation, the angle between successive planes being $\frac{\pi}{2t}$. The wedges formed by the pairs of successive planes will all have the same component of attraction along $\overrightarrow{AC}$. Each wedge is then to be sliced up by $t+1$ lines through A (corresponding to AM, Am), the angle between successive lines being $\frac{\pi}{2t}$, and the corresponding pyramids are to be constructed as before. The whole spheroid is now the union of the pyramids formed in this way. For a particular pair AM, Am take the corresponding pyramid from each wedge. The sum of their individual components of attraction at A in

---

[71]This constant depends on the gravitational constant, the density of the material and the units used.

[72]The *arc XN* has length

$$\rho \cos^{-1} \left(\frac{\cos \alpha}{\sqrt{\sin^2 \theta + \cos^2 \alpha \cos^2 \theta}} \right),$$

where $\rho = |AC|$ and $\theta = \angle CAN$. For small α we can show that this is approximately $\rho \tan \alpha \sin \theta \approx |RN| \tan \alpha$.

the direction $\overrightarrow{AC}$ is the attraction at A of the solid obtained by rotating the area AMm about the axis Aa (note that other components will cancel out in pairs); from (2) this is approximately

$$4t \times \frac{k}{\rho}|RK||Rr|\tan\frac{\pi}{2t} = \frac{2k\pi}{\rho}\left(\frac{\tan\frac{\pi}{2t}}{\frac{\pi}{2t}}\right)|RK||Rr|.$$

An approximation to the attraction at A of the whole spheroid is then given by the sum over each pair AM, Am, namely,

$$\frac{2k\pi}{\rho}\left(\frac{\tan\frac{\pi}{2t}}{\frac{\pi}{2t}}\right)\sum|RK||Rr|.$$

In the limit as $t \to \infty$ we should have equality, in which case the attraction at A of the whole spheroid will turn out to be

$$\frac{2k\pi}{\rho}\int_{[A,C]}|RK|\,d|AR| = \frac{2k\pi}{|AC|}\int_{[A,C]}|RK|\,d|AR|. \tag{3}$$

This appears to be the essence of MacLaurin's argument.

Again MacLaurin has evaluated by a series of approximations what is in effect a triple integral. The attraction at the origin A of the solid V formed by rotating about the x-axis the graph $y = f(x)$, where $0 \le x \le 2a$, $f(x) \ge 0$ and $f(0) = f(2a) = 0$, is given by (see equation (5) in Appendix III.3, p. 202)

$$k\iiint_V \frac{x}{(x^2 + y^2 + z^2)^{3/2}}\,dx\,dy\,dz$$

$$= 4k\int_0^{2a}\int_0^{\pi/2}\int_0^{f(x)} \frac{x}{(x^2 + r^2)^{3/2}}r\,dr\,d\theta\,dx$$

$$= 2k\pi\int_0^{2a}\left[\frac{-x}{\sqrt{x^2 + r^2}}\right]_{r=0}^{r=f(x)}dx = 2k\pi\int_0^{2a}1 - \frac{x}{\sqrt{x^2 + (f(x))^2}}\,dx$$

$$= 2k\pi\left(2a - \int_0^{2a}\frac{x}{\sqrt{x^2 + (f(x))^2}}\,dx\right). \tag{4}$$

Now we show that (3) leads to the same expression. For this we assume in addition that the expression $\frac{x}{\sqrt{x^2+(f(x))^2}}$ defines a strictly increasing function on $(0, 2a]$, for which a sufficient condition, obtained from the derivative, is that $xf'(x) < f(x)$ on $(0, 2a)$, and moreover that it has limit 0 as x tends to 0 from the right. These assumptions will be true in the application in Proposition II. Referring to MacLaurin's diagram, we put $|AR| = t$, $|AQ| = x$, $|AC| = \rho$, so that

$$\frac{t}{\rho} = \frac{|AR|}{|AN|} = \frac{|AQ|}{|AM|} = \frac{x}{\sqrt{x^2 + (f(x))^2}} \quad \text{and} \quad \int_{[A,C]}|RK|\,d|AR| = \int_0^\rho x\,dt.$$

Now we apply the formula[73]

$$\int_0^\alpha h(u)\,du + \int_0^\beta h^{-1}(v)\,dv = \alpha\beta,$$

where h is a continuous, strictly increasing function on $[0,\alpha]$ with $h(0) = 0$ and $h(\alpha) = \beta$, to get

$$\int_0^{2a} t\,dx + \int_0^\rho x\,dt = 2a\rho,$$

from which we obtain

$$\int_{[A,C]} |RK|d|AR| = \rho\left(2a - \int_0^{2a} \frac{x}{\sqrt{x^2 + (f(x))^2}}\,dx\right).$$

Substituting this in (3) produces the result in (4).

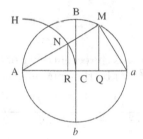

Diagram for the Corollary

The Corollary is concerned with the case where the curve $ABab$ is a circle with centre C. Now we have

$$\frac{|AQ|}{|Aa|} = \frac{|AQ|}{|AM|} \cdot \frac{|AM|}{|Aa|} = \frac{|AQ|}{|AM|} \cdot \frac{|AQ|}{|AM|} = \frac{|AR|^2}{|AN|^2} = \frac{|AR|^2}{|AC|^2},$$

and so

$$|RK| = |AQ| = \frac{2}{|AC|} \times |AR|^2.$$

Then for the sphere

$$\int_{[A,C]} |RK|\,d|AR| = \frac{2}{|AC|} \int_0^{|AC|} x^2\,dx = \frac{2}{3}|AC|^2,$$

and by (3) the attraction at A (or at any point on the surface) is $\frac{4}{3}k\pi a$. Taking the ratio of the respective quantities given by (3) for the general figure of the Lemma and the sphere with Aa as diameter leads to the stated result.

---

[73]This elementary formula, which is obvious from a diagram, does not seem to be well known nowadays.

Note on Proposition II (pp. 118–119). Here MacLaurin determines the attraction at a pole of a spheroid formed by rotating an ellipse about its major axis (oblong spheroid). His argument begins with two geometrical observations. First, $|AM| = 2|Ce|$ since Ce is parallel to AM and bisects Aa. Less trivial is the assertion that $\frac{|Ce|}{|CV|} = \frac{|CL|}{|Ca|}$. This is easily seen in the case of a circle (see diagram below) because of the right-angles, for then $|CV| = |Ca|$ of course, and $|Ce| = |CL|$ since triangles CLV and Cea are congruent; the result may then be deduced for the ellipse by projection (see Appendix III.1, pp. 197–199).

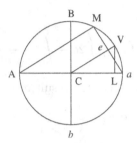

Note that $c = ae$, where e is the eccentricity of the ellipse, and $c^2 = a^2 - b^2$; also, $|VL|^2 = b^2(1 - |CL|^2/a^2)$ by the canonical equation of the ellipse. From

$$\frac{x^2}{a^2 - x^2} = \frac{l^2}{(a^2 - l^2) \times \frac{b^2}{a^2}} \tag{1}$$

MacLaurin obtains

$$|RK| = |AQ| = \frac{2l^2}{a} = \frac{2ab^2x^2}{a^4 - c^2x^2},$$

so that in the notation of NLV (equation (3), p. 168)

$$\int_{[A,C]} |RK|\, d|AR| = 2ab^2 \int_0^a \frac{x^2}{a^4 - c^2x^2}\, dx \qquad (\text{put } z = \tfrac{c}{a}x,\ dz = \tfrac{c}{a}dx)$$

$$= \frac{2a^2b^2}{c^3} \int_0^c \frac{z^2}{a^2 - z^2}\, dz = \frac{2a^2b^2}{c^3} \int_0^c -1 + \frac{a}{2}\left(\frac{1}{a - z} + \frac{1}{a + z}\right) dz$$

$$= \frac{2a^2b^2}{c^3} \left[-z + \frac{a}{2}\ln\frac{a + z}{a - z}\right]_0^c = \frac{2a^2b^2}{c^3}\left(a\ln\sqrt{\frac{a + c}{a - c}} - c\right). \tag{2}$$

To obtain the result given by MacLaurin we have to note that by the logarithm of a quantity t in the system with modulus a MacLaurin means the quantity $a\ln(t/a)$, so he has

$$\ell = a\ln\sqrt{\frac{a + z}{a - z}} \quad \text{and} \quad \mathcal{L} = a\ln\sqrt{\frac{a + c}{a - c}} = a\ln\frac{a + c}{\sqrt{a^2 - c^2}} = a\ln\frac{a + c}{b}.$$

Finally, by the Corollary to Lemma 5 we have

$$\frac{|\text{Attraction at } A \text{ of spheroid}|}{|\text{Attraction at } A \text{ of sphere}|} = \frac{2a^2b^2}{c^3}(\mathcal{L} - c) \times \frac{1}{\frac{2}{3}a^2} = \frac{3b^2}{c^3}(\mathcal{L} - c). \quad (3)$$

MacLaurin's use of calculus and the integral sign in his proof of Proposition II (see also Proposition III (p. 121) and *Remarks* II (pp. 135–136)) is striking. As pointed out by Todhunter, the corresponding parts of [69] use geometric and fluxional arguments; perhaps MacLaurin felt that the version presented in the essay would be more in tune with continental tastes.

In the Scholium MacLaurin states corresponding results for the attraction at the pole of an oblate spheroid. His discussion will be found in Article 646 of [69]. In fact we just follow the same steps with a and b interchanged, which brings about a significant change in the integral to be evaluated. We now have in place of (1)

$$\frac{x^2}{b^2 - x^2} = \frac{\ell^2}{(b^2 - l^2) \times \frac{a^2}{b^2}},$$

which leads to

$$\ell^2 = \frac{a^2b^2x^2}{b^4 + c^2x^2},$$

where we have used $a^2 = b^2 + c^2$, and consequently

$$|RK| = \frac{2}{b} \times \frac{a^2b^2x^2}{b^4 + c^2x^2} = \frac{2a^2bx^2}{b^4 + c^2x^2}.$$

Then

$$\int_{[B,C]} |RK| \, d|AR| = 2a^2b \int_0^b \frac{x^2}{b^4 + c^2x^2} \, dx \qquad (\text{put } z = \tfrac{c}{b}x, \, dz = \tfrac{c}{b}dx)$$

$$= \frac{2a^2b^2}{c^3} \int_0^c \frac{z^2}{b^2 + z^2} \, dz = \frac{2a^2b^2}{c^3} \int_0^c 1 - \frac{b^2}{b^2 + z^2} \, dz$$

$$= \frac{2a^2b^2}{c^3} \left[z - b\tan^{-1}\frac{z}{b} \right]_0^c = \frac{2a^2b^2}{c^3} \left(c - b\tan^{-1}\frac{c}{b} \right).$$

The required ratio is then the result of dividing this by $\frac{2}{3}b^2$ (Corollary to Lemma V), namely,

$$\frac{3a^2}{c^3} \left(c - b\tan^{-1}\frac{c}{b} \right). \quad (4)$$

MacLaurin expresses this in terms of lengths. Referring to his Fig. 7 (pp. 115, 157), we see that $\angle CBF$ has tangent $\frac{|CF|}{|BC|} = \frac{c}{b}$, so $b\tan^{-1}\frac{c}{b}$ is the length of the circular arc CS. We may therefore write (4) as

$$\frac{3|CA|^2}{|CF|^3} \left(|CF| - |\text{arc } CS| \right). \quad (5)$$

Newton dealt with the attraction of a spheroid at any point on its axis produced in Corollary 2 of Proposition XCI of the *Principia* [85] ([15, 63]).

Note on Lemma VI (pp. 119–120). Lemma VI does not appear to apply as generally as Lemma V, so let us assume in this discussion that we are dealing with the situation in the intended application of Proposition III, where we have a spheroid through which we are slicing with planes normal to its equatorial plane and containing a given point on its equator. The section of the spheroidal surface by such a plane is an ellipse; moreover, this ellipse is always similar to the generating ellipse, a fact which is important in the intended application (see Appendix III.2, pp. 199–201). Note that the ellipse $BAba$ shown in MacLaurin's diagram below is the section by such a plane, and is not necessarily the generating ellipse.

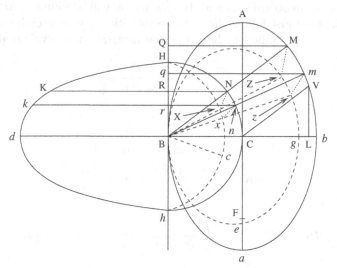

Diagram for Lemma VI and Prop. III
MacLaurin's Fig. 9

Exactly as in Lemma V (see NLV, p. 167) we obtain the approximation

$$\frac{k}{|BC|}|RK||Rr|\tan\alpha$$

for the component in the direction $\overrightarrow{Bb}$ of the attraction at B of the pyramid $BMZzm$; the quantities k and α are as before and

$$\cos\angle bBM = \sin\angle MBQ = \sin\angle NBR = \frac{|NR|}{|BC|}.$$

Now let $\alpha = \pi/t$, where t is an arbitrary positive integer, and partition the wedge between the planes by taking $t+1$ half-lines from B (such as BM, Bm), the angle between successive lines being π/t, and constructing the corresponding pyramids. We obtain as an approximation to the magnitude of the attraction at B from the spheroidal wedge

$$\frac{k}{|BC|}\tan\frac{\pi}{t}\sum|RK||Rr|,\tag{1}$$

and as an approximation to the magnitude of the attraction at B from the spherical wedge

$$\frac{k}{|BC|}\tan\frac{\pi}{t}\sum|RN||Rr|.\tag{2}$$

The ratio of the quantity in (1) to the quantity in (2), that is to say, $\sum|RK||Rr|/\sum|RN||Rr|$, has limit as $t\to\infty$

$$\frac{\int_{[H,h]}|RK|\,d|HR|}{\int_{[H,h]}|RN|\,d|HR|}.\tag{3}$$

The denominator is just the area of the semicircle with radius $|BC|$, that is, $\frac{1}{2}\pi|BC|^2$. See also the further remarks in NPIII (p. 176) on the interpretation of these integrals.

We note for the application in Proposition III that this limit is independent of the particular sectioning plane, since we always have similar sections. The direction $\vec{Bb}$, however, does depend on the plane, but by the symmetry of the spheroid the attraction at B will be perpendicular to the axis, so we need only consider the components perpendicular to the axis of the two attractions above, which will still have the same limiting ratio as that given in (3).

The Corollary is concerned with the case where the generating ellipse is a circle.

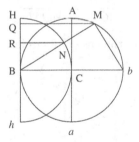

Diagram for the Corollary

We now have

$$\frac{|QM|}{|RN|}=\frac{|BM|}{|BN|}=\frac{|BM|}{|Bb|}\times\frac{|Bb|}{|BN|}=\frac{|RN|}{|BN|}\times\frac{|Bb|}{|BN|},$$

so that

$$\frac{|QM|}{|Bb|}=\frac{|RN|^2}{|BN|^2}=\frac{|RN|^2}{|BC|^2}$$

and

$$|RK|=|QM|=\frac{2|RN|^2}{|BC|}=\frac{2(|BC|^2-|RB|^2)}{|BC|}.$$

Then

$$\int_{[H,h]} |RK| \, d|HR| = \frac{2}{|BC|} \int_{-|BC|}^{|BC|} |BC|^2 - x^2 \, dx$$

$$= \frac{2}{|BC|} \left[|BC|^2 x - \frac{x^3}{3} \right]_{-|BC|}^{|BC|} = \frac{8}{3} |BC|^2 \,.$$

Now suppose we use two particular planes as above to form a wedge of the spheroid and a wedge of the sphere with the same equator. Then, applying (3) to both wedges, we see that, when we let $\alpha \to 0$ keeping one of the planes fixed, the limiting value of

$$\frac{|\text{Attraction at } B \text{ of spheroidal wedge}|}{|\text{Attraction at } B \text{ of spherical wedge}|}$$

is the ratio of the quantity $\int_{[H,h]} |RK| \, d|HR|$ for the spheroidal wedge to $\frac{8}{3}|BC|^2$, where C is the centre for the fixed plane. Again this limit is independent of the particular sectioning plane, and we can replace each attraction by its component perpendicular to the axis of the spheroid.

Note on Proposition III (pp. 120–122). The argument of Proposition III is similar to that of Proposition II. Again MacLaurin uses without comment a result concerning the ellipse which perhaps requires some justification:

$$\frac{|CB|}{|CL|} = \frac{|CL|}{\frac{1}{2}|QM|} \,.$$

This is easily established in the case of a circle:

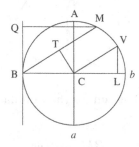

if T is the midpoint of BM we have that triangles BTC and CLV are congruent and

$$\frac{|CB|}{|CL|} = \frac{|CV|}{|CL|} = \frac{|BM|}{|QM|} = \frac{|BT|}{\frac{1}{2}|QM|} = \frac{|CL|}{\frac{1}{2}|QM|} \,.$$

The result for the ellipse can now be deduced by projection (see Appendix III.1, pp. 197–199). Then we obtain easily

$$|RK| = |QM| = \frac{2l^2}{b} = \frac{2a^2b(b^2 - x^2)}{a^2b^2 - c^2x^2},$$

and so

$$\int_{[H,h]} |RK|\, d|HR| = 2a^2b \int_{-b}^{b} \frac{b^2 - x^2}{a^2b^2 - c^2x^2}\, dx \qquad \text{(put } z = \tfrac{c}{b}x,\ dz = \tfrac{c}{b}dx\text{)}$$

$$= \frac{2a^2b^2}{c^3} \int_{-c}^{c} \frac{c^2 - z^2}{a^2 - z^2}\, dz = \frac{4a^2b^2}{c^3} \int_{0}^{c} 1 + \frac{c^2 - a^2}{2a}\left(\frac{1}{a - z} + \frac{1}{a + z}\right) dz$$

$$= \frac{4a^2b^2}{c^3}\left[z - \frac{b^2}{2a}\ln\frac{a + z}{a - z}\right]_0^c = \frac{4a^2b^2}{c^3}\left(c - \frac{b^2}{2a}\ln\frac{a + c}{a - c}\right)$$

$$= \frac{4b^2}{c^3}\left(a^2c - b^2\mathcal{L}\right),$$

where $\mathcal{L} = a\ln\sqrt{\frac{a+c}{a-c}}$ as before.

Now we partition the spheroid by means of a succession of $t + 1$ half-planes, each normal to the equatorial plane and containing the point B; the angle between successive planes is π/t and the first and last form the tangent plane to the spheroid at B. The attraction at B of the whole spheroid is the vector sum of the attractions at B of each of the wedges formed by successive planes and, because the direction of the attraction must be perpendicular to the axis by symmetry, this is the sum of the components perpendicular to the axis of those attractions. Next we note that for large t, according to the Corollary to Lemma VI, any one of these components is approximately equal to the same component of the attraction at B from the corresponding wedge of the sphere with the same equator, multiplied by the invariant ratio of the quantity $\int_{[H,h]} |RK|\, d|HR|$ to $\frac{8}{3}|BC|^2$. Adding these we get the magnitude of the attraction at B of the sphere multiplied by the same invariant ratio. Thus in the limit we should have

$$\frac{|\text{Attraction at } B \text{ of spheroid}|}{|\text{Attraction at } B \text{ of sphere}|} = \frac{\int_{[H,h]} |RK|\, d|HR|}{\frac{8}{3}|BC|^2} = \frac{\frac{4b^2}{c^3}(a^2c - b^2\mathcal{L})}{\frac{8}{3}b^2}$$

$$= \frac{a^2c - b^2\mathcal{L}}{\frac{2}{3}c^3}. \tag{1}$$

Here, a, b, c can be the quantites corresponding to any of the plane sections considered, in particular the section through B containing the axis, for which a, b, c are the appropriate parameters for the generating ellipse.[74]

---

[74]Note that the final expression in (1) is indeed unaffected if we replace a, b, c by λa, λb, λc, respectively ($\lambda \neq 0$).

The Corollary is concerned with the attraction at B of the portions of the spheroid and of the sphere which lie on the same side of a sectioning half-plane of the above type. It is established in the same way as the Proposition by considering the wedges that make up the relevant parts.

Let us justify briefly MacLaurin's procedure by means of equation (6) of Appendix III.3 (p. 202) (cf. discussion in NLIV, pp. 154–156). We are concerned with the spheroid with equation $\frac{x^2}{b^2} + \frac{y^2}{b^2} + \frac{z^2}{a^2} = 1$ and we want the attraction at a point on the equator; by symmetry it is enough to consider $(-b, 0, 0)$. The attraction is given by

$$k \int_0^\pi \int_0^\pi \int_0^{\rho(\theta,\phi)} \sin^2 \phi \cos \theta \, d\rho \, d\phi \, d\theta, \qquad (*)$$

where $\rho(\theta, \phi)$ turns out to be

$$\frac{2a^2 b \sin \phi \cos \theta}{a^2 \sin^2 \phi + b^2 \cos^2 \phi}.$$

We therefore have

$$k \int_0^\pi \int_0^\pi \frac{2a^2 b \sin^3 \phi \cos^2 \theta}{a^2 \sin^2 \phi + b^2 \cos^2 \phi} \, d\phi \, d\theta$$
$$= k \int_0^\pi \frac{2a^2 b \sin^3 \phi}{a^2 \sin^2 \phi + b^2 \cos^2 \phi} \, d\phi \int_0^\pi \cos^2 \theta \, d\theta$$
$$= k\pi a^2 b \int_0^\pi \frac{\sin^3 \phi}{a^2 \sin^2 \phi + b^2 \cos^2 \phi} \, d\phi.$$

For $\int_{[H,h]} |RK| \, d|HR|$ above, MacLaurin has to evaluate

$$\int_{-b}^b \frac{b^2 - x^2}{a^2 b^2 - c^2 x^2} \, dx;$$

under the substitution $x = b \cos \phi$ this transforms to a multiple of the previous integral. Effectively, in Lemma VI MacLaurin is just setting up in geometric terms the double integral with respect to ρ and ϕ in $(*)$. Its dependence on θ is the factor $\cos \theta$, which produces the constant factor $\pi/2$ in the final expression; these cancel out when he compares the spheroidal case with the spherical case.

In the Scholium MacLaurin states a corresponding result for the attraction of an oblate spheroid at any point on its equator; MacLaurin's discussion will be found in Article 646 of [69]. Again the analysis is very similar to that for the oblong spheroid except that the roles of a and b are interchanged (cf. discussion of the Scholium in NPII). We sketch some details. The initial relation is now

$$\frac{a^2 - x^2}{x^2} = \frac{l^2}{(a^2 - l^2) \times \frac{b^2}{a^2}},$$

which leads to

$$l^2 = \frac{a^2b^2(a^2 - x^2)}{a^2b^2 + c^2x^2},$$

and therefore

$$|RK| = |MQ| = \frac{2l^2}{a} = \frac{2ab^2(a^2 - x^2)}{a^2b^2 + c^2x^2}.$$

Then

$$\int_{[H,h]} |RK|\, d|HR| = 2ab^2 \int_{-a}^{a} \frac{a^2 - x^2}{a^2b^2 + c^2x^2}\, dx \qquad (\text{put } z = \tfrac{c}{a}x,\ dz = \tfrac{c}{a}dx)$$

$$= \frac{2a^2b^2}{c^3} \int_{-c}^{c} \frac{c^2 - z^2}{b^2 + z^2}\, dz = \frac{4a^2b^2}{c^3} \int_0^c -1 + \frac{b^2 + c^2}{b^2 + z^2}\, dz$$

$$= \frac{4a^2b^2}{c^3} \left[-z + \frac{a^2}{b} \tan^{-1} \frac{z}{b} \right]_0^c = \frac{4a^2b^2}{c^3} \left(\frac{a^2}{b} \tan^{-1} \frac{c}{b} - c \right).$$

The analogue of (1) then turns out to be[75]

$$\frac{\frac{4a^2b^2}{c^3} \left(\frac{a^2}{b} \tan^{-1} \frac{c}{b} - c \right)}{\frac{8}{3}a^2} = \frac{|CA|^2 |\text{arc } CS| - |CB|^2 |CF|}{\frac{2}{3}|CF|^3}. \qquad (2)$$

MacLaurin notes in Article 647 of [69] that the determination of the attraction at the equator was first resolved by Stirling (see also the MacLaurin - Stirling correspondence, especially the letter from MacLaurin dated 6 December 1740 ([77], Letter 171; [111], pp. 90–92)).

Note on Proposition IV (pp. 122–124). MacLaurin makes use of three ratios which he has already determined:

(i)   (Proposition II; see equation (3) of NPII, p. 171)

$$\frac{|\text{Attraction at } A \text{ of spheroid}|}{|\text{Attraction at } A \text{ of sphere with diameter } Aa|} = \frac{3b^2(\mathcal{L} - c)}{c^3};$$

(ii) (Corollary 1 of Lemma III; see equation (1) of NLIII, p. 147)

$$\frac{|\text{Attraction at } A \text{ of sphere with diameter } Aa|}{|\text{Attraction at } B \text{ of sphere with diameter } Bb|} = \frac{a}{b};$$

(iii) (Proposition III; see equation (1) of NPIII, p. 175)

$$\frac{|\text{Attraction at } B \text{ of sphere with diameter } Bb|}{|\text{Attraction at } B \text{ of spheroid}|} = \frac{\frac{2}{3}c^3}{a^2c - b^2\mathcal{L}}.$$

---

[75]See MacLaurin's Fig. 7 (pp. 115, 157) and NPII (p. 171).

Multiplying these together gives

$$\frac{\mathcal{A}}{\mathcal{B}} = \frac{|\text{Attraction at } A \text{ of spheroid}|}{|\text{Attraction at } B \text{ of spheroid}|} = \frac{2ab(\mathcal{L} - c)}{a^2c - b^2\mathcal{L}} . \tag{1}$$

Since we are treating the Earth as a fluid spheroid in equilibrium under the attractions of the Moon and the Sun, the conditions of Proposition I must hold. Thus we have[76]

$$\mathcal{A}a - \mathcal{B}b = \frac{2a^2V + b^2V}{d} . \tag{2}$$

Substituting in this for $\mathcal{B}$ from (1) leads to

$$\frac{V}{\mathcal{A}} = \frac{2a^2\mathcal{L} + b^2\mathcal{L} - 3a^2c}{\frac{2a}{d}(2a^2 + b^2)(\mathcal{L} - c)} . \tag{3}$$

Now, since $0 \le \frac{c}{a} < 1$, we have

$$\mathcal{L} = \frac{a}{2} \ln \frac{a+c}{a-c} = \frac{a}{2} \ln \frac{1 + \frac{c}{a}}{1 - \frac{c}{a}} = a \sum_{n=0}^{\infty} \frac{c^{2n+1}}{(2n+1)a^{2n+1}} ,$$

and therefore

$$\mathcal{L} - c = a \sum_{n=1}^{\infty} \frac{c^{2n+1}}{(2n+1)a^{2n+1}} . \tag{4}$$

MacLaurin uses the series for $\mathcal{L}$ to derive further versions of equations (1) and (3). For the numerator in (3) we have

$$2a^2\mathcal{L} + b^2\mathcal{L} - 3a^2c = (3a^2 - c^2)\mathcal{L} - 3a^2c$$

$$= 3a^3 \sum_{n=1}^{\infty} \frac{c^{2n+1}}{(2n+1)a^{2n+1}} - ac^2 \sum_{n=0}^{\infty} \frac{c^{2n+1}}{(2n+1)a^{2n+1}}$$

$$= 3c^3 \sum_{n=1}^{\infty} \frac{c^{2n-2}}{(2n+1)a^{2n-2}} - c^3 \sum_{n=0}^{\infty} \frac{c^{2n}}{(2n+1)a^{2n}}$$

$$= c^3 \sum_{n=1}^{\infty} \left(\frac{3}{2n+1} - \frac{1}{2n-1} \right) \left(\frac{c}{a} \right)^{2n-2}$$

$$= 4c^3 \sum_{n=2}^{\infty} \frac{n-1}{4n^2 - 1} \left(\frac{c}{a} \right)^{2n-2} = 4c^3 \sum_{n=1}^{\infty} \frac{n}{(2n+1)(2n+3)} \left(\frac{c}{a} \right)^{2n} ,$$

and so[77]

---

[76]See NPI, equation (6), and subsequent comments in relation to Corollaries 4 and 5 (pp. 163–164).

[77]The denominator is given incorrectly in [68].

$$\frac{V}{\mathcal{A}} = \frac{2\displaystyle\sum_{n=1}^{\infty} \frac{n}{(2n+1)(2n+3)}\left(\frac{c}{a}\right)^{2n}}{\dfrac{a}{c^3 d}(2a^2+b^2)(\mathcal{L}-c)}.$$

(5)

Using (1) we deduce from (5)

$$\frac{V}{\mathcal{B}} = \frac{2\displaystyle\sum_{n=1}^{\infty} \frac{n}{(2n+1)(2n+3)}\left(\frac{c}{a}\right)^{2n}}{\dfrac{1}{2bc^3 d}(2a^2+b^2)(a^2c-b^2\mathcal{L})},$$

(6)

and then from (5) and (6) we obtain for $\frac{1}{2}(\mathcal{A}+\mathcal{B})/V$ the expression

$$\left(4\sum_{n=1}^{\infty} \frac{n}{(2n+1)(2n+3)}\left(\frac{c}{a}\right)^{2n}\right)^{-1} \frac{2a^2+b^2}{c^3 d}\left(a(\mathcal{L}-c)+\frac{1}{2b}(a^2c-b^2\mathcal{L})\right),$$

which simplifies to give MacLaurin's expression

$$\frac{V}{\frac{1}{2}(\mathcal{A}+\mathcal{B})} = \frac{2\displaystyle\sum_{n=1}^{\infty} \frac{n}{(2n+1)(2n+3)}\left(\frac{c}{a}\right)^{2n}}{\dfrac{2a^2+b^2}{4bc^3 d}\left(2ab\mathcal{L}-b^2\mathcal{L}+a^2c-2abc\right)}.$$

For the denominator in (1) we have

$$a^2c - b^2\mathcal{L} = a^2c - ab^2\sum_{n=0}^{\infty} \frac{c^{2n+1}}{(2n+1)a^{2n+1}}$$

$$= (a^2-b^2)c - a(a^2-c^2)\sum_{n=1}^{\infty} \frac{c^{2n+1}}{(2n+1)a^{2n+1}}$$

$$= c^3 - \sum_{n=1}^{\infty} \frac{c^{2n+1}}{(2n+1)a^{2n-2}} + \sum_{n=1}^{\infty} \frac{c^{2n+3}}{(2n+1)a^{2n}}$$

$$= \sum_{n=0}^{\infty}\left(\frac{1}{2n+1}-\frac{1}{2n+3}\right)\frac{c^{2n+3}}{a^{2n}} = 2\sum_{n=0}^{\infty} \frac{c^{2n+3}}{(2n+1)(2n+3)a^{2n}}.$$

(7)

Then from (1), (4) and (7) we obtain

$$\frac{A}{B} = \frac{a^2 b \displaystyle\sum_{n=1}^{\infty} \frac{c^{2n+1}}{(2n+1)a^{2n+1}}}{\displaystyle\sum_{n=0}^{\infty} \frac{c^{2n+3}}{(2n+1)(2n+3)a^{2n}}} = \frac{\dfrac{b}{a} \displaystyle\sum_{n=1}^{\infty} \frac{c^{2n-2}}{(2n+1)a^{2n-2}}}{\displaystyle\sum_{n=0}^{\infty} \frac{c^{2n}}{(2n+1)(2n+3)a^{2n}}}$$

$$= \frac{b \displaystyle\sum_{n=0}^{\infty} \frac{c^{2n}}{(2n+3)a^{2n}}}{a \displaystyle\sum_{n=0}^{\infty} \frac{c^{2n}}{(2n+1)(2n+3)a^{2n}}} ,$$

as given by MacLaurin. For small c this is approximately

$$\frac{b\left(\frac{1}{3} + \frac{c^2}{5a^2}\right)}{a\left(\frac{1}{3} + \frac{c^2}{15a^2}\right)} ,$$

and if we write $a = d + x$, $b = d - x$, so that $c^2 = (d+x)^2 - (d-x)^2 = 4xd$, this becomes

$$\frac{(d-x)\left(\frac{1}{3} + \frac{4x}{5d}\left(1 + \frac{x}{d}\right)^{-2}\right)}{(d+x)\left(\frac{1}{3} + \frac{4x}{15d}\left(1 + \frac{x}{d}\right)^{-2}\right)} \approx \frac{(d-x)\left(\frac{1}{3} + \frac{4x}{5d}\right)}{(d+x)\left(\frac{1}{3} + \frac{4x}{15d}\right)}$$

$$\approx \frac{\frac{1}{3}d + \left(\frac{4}{5} - \frac{1}{3}\right)x}{\frac{1}{3}d + \left(\frac{4}{15} + \frac{1}{3}\right)x} = \frac{\frac{1}{3}d + \frac{7}{15}x}{\frac{1}{3}d + \frac{9}{15}x} . \tag{8}$$

This is not the approximation given by MacLaurin, namely, $\frac{1}{3}d + \frac{17}{15}x : \frac{1}{3}d + \frac{19}{15}x$, which I believe is wrong. MacLaurin continues his investigations by applying his ratio to determine $\frac{x}{d}$. However, he takes up the matter again in *Remarks* I at the end of his dissertation (p. 133), where he obtains an improved version and refers to an earlier "error of the pen or calculation." I will therefore use (8) in the subsequent discussion; it does give values which are closer to those obtained in the *Remarks*.

Taking $\dfrac{A}{B} = \dfrac{\frac{1}{3}d + \frac{7}{15}x}{\frac{1}{3}d + \frac{9}{15}x}$ leads to

$$\frac{B - A}{G} = \frac{B - A}{\frac{1}{2}(A + B)} = \frac{\frac{2}{15}x}{\frac{1}{3}d + \frac{8}{15}x} = \frac{2x}{5d + 8x} . \tag{9}$$

Now, according to Corollary 2 of Proposition I (see NPI, p. 163),

$$\frac{x}{d} \approx \frac{B - A + 3V}{B + A - 2V} ,$$

which, on substitution from (9), produces

$$\frac{x}{d} \approx \frac{\frac{2Gx}{5d+8x}+3V}{2G-2V} = \frac{2Gx+15Vd+24Vx}{2(G-V)(5d+8x)},$$

and so

$$2(G-V)(5xd+8x^2) \approx 2Gxd+15Vd^2+24Vxd.$$

Ignoring the term in x^2, which is small, leads to the approximation

$$(8G-34V)xd \approx 15Vd^2,$$

that is to say,

$$\frac{x}{d} \approx \frac{15V}{8G-34V}. \tag{10}$$

Since V is small in comparison with G, we may even take

$$\frac{x}{d} \approx \frac{15V}{8G}, \tag{11}$$

which also follows from MacLaurin's version and is applied by him. We note for the application in Proposition V that the results are valid for the effects of the Sun and the Moon separately with appropriate interpretation of V.

For the Corollary we use (9) and (11):

$$\frac{B-A}{G} \approx \frac{2x}{5d+8x} = \frac{2x}{5d}\left(1+\frac{8x}{5d}\right)^{-1} \approx \frac{2x}{5d} \approx \frac{2}{5} \times \frac{15V}{8G} = \frac{3V}{4G},$$

so

$$B-A \approx \frac{3V}{4} \quad \text{and} \quad B-G = \frac{1}{2}(B-A) \approx \frac{3V}{8}.$$

The result stated in the Scholium concerning an oblate spheroid is established by multiplying together the following three ratios:

(i)$'$ (Scholion to Proposition II; see equation (5) of NPII, p. 171)

$$\frac{|\text{Attraction at pole } B \text{ of spheroid}|}{|\text{Attraction at } B \text{ of sphere with diameter } Bb|}$$

$$= \frac{3|CA|^2}{|CF|^3}(|CF| - |\text{arc } CS|);$$

(ii)$'$ (Corollary 1 of Lemma III; see equation (1) of NLIII, p. 147)

$$\frac{|\text{Attraction at } B \text{ of sphere with diameter } Bb|}{|\text{Attraction at } A \text{ of sphere with diameter } Aa|} = \frac{|CB|}{|CA|};$$

(iii)$'$ (Scholium to Proposition III; see equation (2) of NPIII, p. 177)

$$\frac{|\text{Attraction at } A \text{ of sphere with diameter } Aa|}{|\text{Attraction at } A \text{ of spheroid}|}$$

$$= \frac{\frac{2}{3}|CF|^3}{|CA|^2|\text{arc } CS| - |CB|^2|CF|}.$$

The result is established in Article 646 of [69].

Note on Proposition V (pp. 125–127). Here MacLaurin's chief purpose is to use his results in conjunction with values given in Newton's *Principia* to calculate the quantity $2x$, the maximum difference in the heights of the water due to the effect of the Sun. Certain formulae are stated which perhaps require some explanation. First there is

$$\frac{v}{K} = \frac{dT}{ST}.$$ (1)

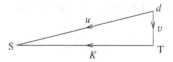

Referring to the above diagram, we have

$$v = u\sin \angle TSd \approx u\tan \angle TSd \approx K\tan \angle TSd = K \times \frac{|dT|}{|ST|}.$$

Next MacLaurin requires

$$\frac{K}{g} = \left(\frac{ST}{S^2}\right) / \left(\frac{dT}{l^2}\right).$$ (2)

If a particle of mass m is moving with uniform angular velocity ω in a circle of radius r under a centripetal force with magnitude F, then the period of the motion is $\tau = 2\pi/\omega$ and $F = mr\omega^2$. Thus $F = 4\pi^2 mr/\tau^2$ and the force per unit mass is $4\pi^2 r/\tau^2$. Applying this to the motion of the Earth about the Sun and the motion of the Moon about the Earth, we obtain

$$\frac{K}{g} \approx \frac{4\pi^2 |ST|}{S^2} \times \frac{l^2}{4\pi^2 |dT|} = \frac{|ST|}{S^2} \times \frac{l^2}{|dT|}.$$

Then

$$\frac{v}{g} = \frac{l^2}{S^2} \leq \frac{L^2}{S^2}$$ (3)

follows on multiplying (1) and (2) and noting that $l^2 \leq L^2$.

The values

$$\frac{L^2}{S^2} = \frac{1}{178\frac{29}{40}} = \frac{1}{178.725} \quad \text{and} \quad \frac{V}{v} = \frac{1}{60\frac{1}{2}}$$

are given by Newton in Propositions XXV and XXXVI, respectively, of Book III of the *Principia* (see [85] ([15, 63])). MacLaurin also requires the value of g/G but does does not give this explicitly; Newton has

$$\frac{v}{G} = \frac{60\frac{1}{2}}{60 \times 60 \times 60 \times 178\frac{29}{40}} = \frac{1}{638092.6}$$

(Book III, Proposition XXV [85] ([15, 63])), which, combined with (3), gives

$$\frac{g}{G} = \frac{g}{v} \times \frac{v}{G} \approx 178\frac{29}{40} \times \frac{60\frac{1}{2}}{60 \times 60 \times 60 \times 178\frac{29}{40}} = \frac{60\frac{1}{2}}{60 \times 60 \times 60}.$$

Then he calculates V/G apparently by

$$\frac{V}{G} = \frac{V}{v} \times \frac{v}{g} \times \frac{g}{G} = \frac{1}{60\frac{1}{2}} \times \frac{1}{178.725} \times \frac{60\frac{1}{2}}{60 \times 60 \times 60} = \frac{1}{38604600},$$

which is the value given by Newton in Proposition XXXVI of Book III of the *Principia* [85] ([15, 63]).

MacLaurin now uses this value of V/G to calculate x/d from the expression given in Corollary 2 of Proposition I and developed in Proposition IV (see NPI (p. 163) and NPIV (p. 180–181)):[78]

$$\frac{x}{d} = \frac{15V}{8G - 57\frac{5}{14}V} = \frac{15}{8\frac{G}{V} - 57\frac{5}{14}}.$$

However, as already noted, MacLaurin corrects this to

$$\frac{x}{d} = \frac{15V}{8G - \frac{88}{7}V} = \frac{15}{8\frac{G}{V} - \frac{88}{7}} \tag{4}$$

in *Remarks* I at the end of the dissertation (pp. 133–134). With[79] $d = 19615800$ we obtain from (4)

$$2x = 1.90545\ldots \approx 1\frac{90545}{100000} \approx 1'10\frac{8654}{10000}'',$$

which is the same as the value given by MacLaurin.

An attempt is now made to allow for the fact that the *solid* Earth is not a sphere. According to Newton (Proposition XIX of Book III [85] ([15, 63])), if a, b denote the equatorial and polar radii, respectively, then $\frac{a}{b} = \frac{230}{229}$ and so

$$\alpha = \frac{a}{d} = \frac{2a}{a+b} = \frac{2}{1+\frac{b}{a}} = \frac{460}{459},$$

as MacLaurin states. Accordingly, the expression for x/d given in his *Remarks* is to be replaced by

$$\frac{x}{\alpha d} = \frac{15\alpha V}{8\frac{G}{\alpha} - \frac{88}{7}\alpha V}, \quad \text{giving} \quad \frac{x}{d} = \alpha^3 \times \frac{15}{8\frac{G}{V} - \frac{88}{7}\alpha^2} \approx \alpha^3 \times \frac{15}{8\frac{G}{V} - \frac{88}{7}};$$

---

[78]In the original, MacLaurin cites Cor. 2 of Prop. III here, which is clearly wrong.
[79]This measurement in Parisian feet, which is also used by Newton, is due to Picard (see [103], Article 32).

consequently, the previous value obtained for $2x$ has to be multiplied by α^3, producing

$$2x \approx 1.91793\ldots \approx 1\tfrac{9179}{10000} \approx 1'\,11\tfrac{152}{10000}'' \approx 1'\,11\tfrac{1}{60}''\,,$$

as given by MacLaurin. The final reference is to the expedition to Lapland (1736–1737) which the Royal Academy of Sciences sent out under Maupertuis to determine the length of the arc corresponding to a degree of latitude in the region of the Arctic circle (see [103], Chapter VII and [75]).

In Scholium 1 MacLaurin compares the result of Proposition V with what would have arisen from the hypothesis that $\mathcal{A} = \mathcal{B} = G$. From Corollary 1 of Proposition I (see NPI, p. 163) we would then have

$$\frac{a}{b} = \frac{G + \frac{bV}{d}}{G - \frac{2aV}{d}} \approx \frac{G + V}{G - 2V}\,,$$

and so

$$aG - 2aV \approx bG + bV\,, \quad \text{hence} \quad (a - b)G \approx (2a + b)V\,,$$

from which it follows that

$$2x = a - b \approx (2a + b)\frac{V}{G} \approx \frac{3Vd}{G}\,. \tag{5}$$

Equation (4) gives

$$2x \approx \frac{15Vd}{4G}\,, \tag{6}$$

which exceeds the approximation in (5) by $\frac{3Vd}{4G}$. Scholium 2 refers to Jupiter's spots and contains some speculations about tides on Jupiter caused by its Moons (see [72] and Part I, Note on Proposition VII, p. 27).

Note on the Preamble of Section IV (pp. 127–128). Here MacLaurin appears to be using the material from Newton (p. 102) and the ideas of Corollary 4 of Proposition I (see Notes pp. 138–141, 164). At F there are two forces caused by the combined effects of the Moon and the Sun, one acting along $\overrightarrow{FT}$ with magnitude approximately proportional to $|FT|$, the other acting parallel to $\overrightarrow{aA}$ with magnitude approximately proportional to $3|Ff|$. If V denotes the magnitude of the resultant force at the mean distance d as in Corollary 4 of Proposition I, then these forces will have approximate magnitudes $|FT| \times \frac{V}{d}$ and $3|Ff| \times \frac{V}{d}$, respectively. The component of the latter force in the direction $\overrightarrow{TF}$ will then be approximately

$$3|Ff| \times \frac{V}{d} \times \frac{|Fz|}{|Ff|} = \frac{3V}{d}|Fz|\,,$$

and since it acts away from T, it will raise the water. Now suppose that F is the position of least height; that is to say, F is an extremity of the conjugate (minor) axis of the figure. Then against this component there is the first force together with the excess of the gravitational attraction $\mathcal{B}$ at F over the mean attraction G. For equilibrium we require, using these approximations,[80]

$$\mathcal{B} - G + \frac{V}{d}|FT| = \frac{3V}{d}|Fz|. \tag{1}$$

Now $\mathcal{B} - G \approx \frac{3V}{8}$ by the Corollary to Proposition IV (see NPIV, p. 181), $|FT| \approx d$ and by similar triangles

$$\frac{|Fz|}{|Ff|} = \frac{|Ff|}{|FT|}.$$

Using these in (1) leads to the approximate relation

$$\frac{3V}{8} + V \approx 3V\frac{|Ff|^2}{|FT|^2},$$

and therefore

$$\sin \angle FTb = \frac{|Ff|}{|FT|} \approx \sqrt{\frac{1}{8} + \frac{1}{3}} = \sqrt{\frac{11}{24}};$$

thus $\angle FTb \approx 42° \, 37'$ as MacLaurin has it.

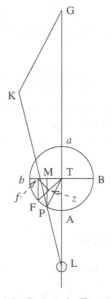

MacLaurin's Fig. 1

---

[80]The two published versions of (1) have $\mathcal{A}$ in place of G. It seems clear from what follows in MacLaurin's text that G is intended.

Note on Proposition VI (pp. 128–130). MacLaurin applies the results of Proposition IV to deduce that the shape taken on by the Moon, supposed to be fluid, under the attraction of the Sun is a spheroid whose generating ellipse is approximately similar to the ellipse in which the Moon orbits the Earth. For the latter, MacLaurin cites observations made by Edmund Halley. In the quadratures the Moon will be approximately at its greatest distance from the Earth, while in the syzygies, assuming the orbit to be nearly circular, the distance of the Moon from the Earth will be approximately the length of the semiminor axis of the orbit. According to Halley's observations, for the lunar orbit[81]

$$\frac{b}{a} = \frac{44.5}{45.5} = 0.9780 \quad \text{and} \quad \frac{a}{b} = 1.0225. \tag{1}$$

MacLaurin also quotes Newton's value of 69/70 for b/a and refers to Newton as "Clarissimus Auctor Tractatus de Motibus Lunae secundum Theoriam gravitatis." The treatise identified is a rare pamphlet of Newton's entitled, *A New and most Accurate Theory of the Moon's Motion*, which was published in 1702; David Gregory included a Latin version in [51]. The pamphlet is reproduced and discussed in [32]. The ratio 70 to 69 appears in it in connection with the horizontal parallax of the Moon in a lunar eclipse (see [32], p. 119).

With the notation and values of Proposition V we have on applying Proposition IV (last line of its proof; see NPIV, pp. 180–181) to the Moon and the Sun[82]

$$\frac{x}{d} = \frac{15v}{8g - 57\frac{5}{14}v} = \frac{15}{8\frac{g}{v} - 57\frac{5}{14}} = \frac{15}{8 \times 178.725 - 57\frac{5}{14}} = \frac{1}{91.496},$$

and so

$$\frac{a}{b} = \frac{d+x}{d-x} = \frac{92.496}{90.496} = 1.0221, \tag{2}$$

which is close to the value in (1) given by Halley's observations. No doubt MacLaurin would have wished to repeat the calculation using the analogue of the corrected formula from *Remarks* I (see pp. 133–134, 190), namely,

$$\frac{x}{d} = \frac{15v}{8g - \frac{88}{7}v},$$

which produces

$$\frac{x}{d} = \frac{1}{94.482} \quad \text{and} \quad \frac{a}{b} = 1.0214. \tag{3}$$

---

[81]Modern tables give $e = 0.0549$ for the eccentricity of the lunar orbit and therefore $\frac{b}{a} = \sqrt{1 - e^2} = 0.9985$ and $\frac{a}{b} = 1.0015$.

[82]The quantity g is "the gravity of the Moon towards the Earth at its mean distance." MacLaurin takes this as the analogue of the Earth's G, presumably since the attractions of the Earth and the Moon must balance at the surface for equilibrium of the fluid.

According to Scholium 1 of Proposition V (see NPV, p. 184), under the hypothesis of uniform gravity over the lunar surface we would have[83]

$$\frac{b}{a} \approx \frac{g-2v}{g+v} = \frac{\frac{g}{v}-2}{\frac{g}{v}+1} = \frac{176.725}{179.725} = 0.9833 \approx \frac{59}{60} \quad \text{and} \quad \frac{a}{b} \approx 1.0170. \quad (4)$$

The disparity in the values in (4) compared with those in (1), (2), and (3) led MacLaurin to conjecture that the orbit is flattened due to a decrease in the attraction of the Earth on the Moon between the syzygies and the quadratures (see Part I, Note on Proposition XVI, p. 29); presumably, the ratio to which he refers is $(\frac{15Vd}{4G})/(\frac{3Vd}{G}) = 5/4$ (see equations (5) and (6) in NPV, p. 184).

The observations at the end of the Proposition concerning the effects of the Earth's rotation on the tides are expanded in Articles 690 and 691 of [69]; MacLaurin finds the height of the water is now approximately two thirds of that previously calculated.

Note on Proposition VII (p. 130). Consider the Earth as a sphere of radius $r = 3{,}956.5$ miles (the average of its polar and equatorial radii) rotating about its axis with constant angular velocity $\pi/12$ radians/hour. The velocity on the surface at latitude $50°$ and the velocity at a position 36 miles further north will have respective magnitudes in miles/hour

$$\frac{\pi r}{12}\cos 50° \quad \text{and} \quad \frac{\pi r}{12}\cos\left(50 + \frac{36\times 180}{\pi r}\right)° ;$$

this represents a decrease at the latter position of

$$\frac{\pi r}{6}\sin\left(50 + \frac{36\times 180}{2\pi r}\right)° \sin\left(\frac{36\times 180}{2\pi r}\right)° \approx 7.25 \,\text{miles/hour} ,$$

which is consistent with MacLaurin's assertion. Generally, in the northern hemisphere, if the tide moves away from the equator, there must be a displacement of the water to the east, since the initial west-east component of velocity of the moving water is greater than the velocity due to the rotation of the Earth at higher latitudes. Correspondingly, if the tide moves water towards the equator, there will be a displacement towards the west.

---

[83]MacLaurin also notes that $\frac{3v}{g} = \frac{1}{59.575} \approx \frac{1}{\frac{1}{2}(59+60)}$. In the original [68] he asserts (in Latin) that for the numbers 59 and 60 the "semidifference is to the semisum as $3v$ to g approximately." I have changed *semidifference* to *difference* in the translation.

Note on Proposition VIII (p. 131). Since the equator $ABab$ is almost circular, g, a focus of the equator, will be close to the centre C as shown in MacLaurin's diagram. The series stated in the Proposition is discussed by MacLaurin in his *Remarks* II (pp. 134–136); it therefore seems preferable to delay our analysis of it until the Note on MacLaurin's Remarks (see pp. 191–195).

Note on Proposition IX (pp. 131–132). Here MacLaurin is largely quoting material contained in Proposition XXXVII of Book III of Newton's *Principia* [85] ([15, 63]). I suspect that the rise of $50\frac{1}{2}$ feet stated by MacLaurin is a typographical error: in Corollary 1 of this Proposition, Newton gives a rise of $10\frac{1}{2}$ feet, which is based on a rise of 1 foot $11\frac{1}{30}$ inches due to the Sun and a corresponding rise of 4.4815 times this brought about by the Moon, namely, 8 feet $7\frac{5}{22}$ inches. Newton makes several references to the tidal measurements of Samuel Sturmy (1633–1669), which were published in the *Philosophical Transactions* [99].[84] MacLaurin's earlier citations of Cassini's observations occur in the three subsections of Section I (see Appendix III.4, pp. 202–205).

The possibility of just one tide in a day at certain places, which MacLaurin justifies in a brief footnote, seems to to be concerned with the following situation.

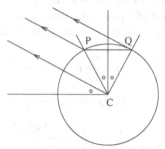

In the above diagram the three indicated angles are all equal to 28°, the angle assumed by MacLaurin for the Moon's declination, and consequently P has latitude 62°. The Moon's attraction acts approximately along the parallel lines indicated and at Q this attraction is perpendicular to the radius CQ. Thus, when the rotation of the Earth moves P round to the position of Q, where it is furthest from the Moon, there is no component of force to raise the water. If the lesser effect of the Sun and other factors affecting tides are ignored, in the simple gravitational model the high tides at a place will normally occur when it is closest to the Moon and also when it is furthest away; thus in the situation described above the second high tide could not occur.

---

[84]See also Sturmy's *The Mariners Magazine* [100]. Chapter 2 of Book I is entitled, "Of the Moons Motion, and the Ebbing and Flowing of the Sea."

Note on MacLaurin's Remarks

Remark I (pp. 133–134). Here MacLaurin returns to the calculation of x/d and gives a corrected and improved version of the result he obtained in Proposition IV (see NPIV, pp. 180–181). He begins with the following formula which was established there:

$$\frac{\mathcal{B}}{\mathcal{A}} = \frac{\frac{1}{3} + \frac{c^2}{15a^2} + \frac{c^4}{35a^4} + \cdots}{\frac{b}{a}\left(\frac{1}{3} + \frac{c^2}{5a^2} + \frac{c^4}{7a^4} + \cdots\right)}.$$

Now

$$\frac{b}{a} = \frac{1}{a}\sqrt{a^2 - c^2} = \left(1 - \frac{c^2}{a^2}\right)^{1/2} = \sum_{n=0}^{\infty}\binom{1/2}{n}\left(-\frac{c^2}{a^2}\right)^n$$

$$= 1 - \frac{c^2}{2a^2} - \frac{c^4}{8a^4} - \cdots,$$

so that, on multiplying the series, we obtain for the denominator

$$\frac{1}{3} + \left(\frac{1}{5} - \frac{1}{6}\right)\frac{c^2}{a^2} + \left(\frac{1}{7} - \frac{1}{10} - \frac{1}{24}\right)\frac{c^4}{a^4} + \cdots = \frac{1}{3} + \frac{c^2}{30a^2} + \frac{c^4}{840a^4} + \cdots.$$

Thus

$$\frac{\mathcal{B}}{\mathcal{A}} = \frac{\frac{1}{3} + \frac{c^2}{15a^2} + \frac{c^4}{35a^4} + \cdots}{\frac{1}{3} + \frac{c^2}{30a^2} + \frac{c^4}{840a^4} + \cdots},$$

and therefore[85]

$$\frac{\mathcal{B} - \mathcal{A}}{\mathcal{G}} = 2 \times \frac{\mathcal{B} - \mathcal{A}}{\mathcal{A} + \mathcal{B}} = 2 \times \frac{\frac{c^2}{30a^2} + \left(\frac{1}{35} - \frac{1}{840}\right)\frac{c^4}{a^4} + \cdots}{\frac{2}{3} + \frac{c^2}{10a^2} + \left(\frac{1}{840} + \frac{1}{35}\right)\frac{c^4}{a^4} + \cdots}$$

$$= \frac{\frac{c^2}{10a^2} + \frac{23c^4}{8 \times 35a^4} + \cdots}{1 + \frac{3c^2}{20a^2} + \frac{25c^4}{8 \times 70a^4} + \cdots}. \tag{1}$$

Next we obtain from[86]

$$a^2 = (d + x)^2 = d^2\left(1 + \frac{2x}{d} + \frac{x^2}{d^2}\right) \quad \text{and} \quad c^2 = 4dx$$

that

$$\frac{c^2}{4a^2} = \frac{x}{d}\left(1 + \frac{2x}{d} + \frac{x^2}{d^2}\right)^{-1} = \frac{x}{d}\left(1 - \left(\frac{2x}{d} + \frac{x^2}{d^2}\right) + \left(\frac{2x}{d} + \frac{x^2}{d^2}\right)^2 - \cdots\right)$$

$$= \frac{x}{d}\left(1 - \frac{2x}{d} - \frac{x^2}{d^2} + \frac{4x^2}{d^2} + \text{terms of higher degree in } x\right)$$

$$= \frac{x}{d} - \frac{2x^2}{d^2} + \frac{3x^3}{d^3} + \cdots.$$

---

[85]In [68] 24 occurs in place of 8 in the numerator of the final expression.
[86]See Proposition IV and NPIV, p. 180.

Substituting this in (1) produces[87]

$$\frac{B-A}{G} = \frac{\frac{1}{10}\left(\frac{4x}{d} - \frac{8x^2}{d^2} + \ldots\right) + \frac{23}{8\times35}\left(\frac{4x}{d} - \ldots\right)^2 + \ldots}{1 + \frac{3}{20}\left(\frac{4x}{d} - \frac{8x^2}{d^2} + \ldots\right) + \frac{25}{8\times70}\left(\frac{4x}{d} - \ldots\right)^2 + \ldots}$$

$$= \frac{\frac{2x}{5d} + \left(\frac{23\times16}{8\times35} - \frac{4}{5}\right)\frac{x^2}{d^2} + \ldots}{1 + \frac{3x}{5d} + \left(\frac{25\times16}{8\times70} - \frac{6}{5}\right)\frac{x^2}{d^2} + \ldots} \approx \frac{\frac{2}{5}xd + \frac{18}{35}x^2}{d^2 + \frac{3}{5}xd - \frac{17}{35}x^2}$$

$$= \frac{14xd + 18x^2}{35d^2 + 21xd - 17x^2}. \tag{2}$$

Now, from Corollary 2 of Proposition I (see equation (7) of NPI, p. 163), we have

$$Bd - Bx + \frac{V}{d}(d^2 - 2dx + x^2) = Ad + Ax - \frac{2V}{d}(d^2 + 2dx + x^2),$$

that is,

$$(B - A)d + 3Vd = 2Gx - 2Vx - \frac{3Vx^2}{d}.$$

Substituting in this for $B - A$ from (2) and ignoring the term involving Vx^2, we obtain the approximate relation

$$\left(\frac{14xd + 18x^2}{35d^2 + 21xd - 17x^2}\right)Gd + 3Vd \approx 2Gx - 2Vx,$$

which leads to

$$x \approx \frac{3 \times 35Vd^2}{56Gd - 133Vd + 24Gx},$$

where the terms in Vx^2, Vx^3 and Gx^3 have been neglected. Finally we use the approximation $\frac{x}{d} \approx \frac{15V}{8G}$ (see Proposition IV and equation (11) of NPIV, p. 181) in the term $24Gx$ to produce MacLaurin's revised expression

$$x \approx \frac{3 \times 35Vd^2}{56Gd - 133Vd + 45Vd} = \frac{3 \times 35Vd}{56G - 88V}. \tag{3}$$

In Article 687 of [69] MacLaurin gives the approximation (in different notation)

$$\frac{x}{d} \approx \frac{15V}{8A - 9\frac{4}{7}V}, \tag{4}$$

which he obtains from the relation (again in different notation)

$$\frac{2a^2 + b^2}{ad} \times \frac{V}{A} = \frac{\frac{2c^2}{5a^2} + \frac{12c^4}{35a^4} + \ldots}{1 + \frac{3c^2}{5a^2} + \ldots}. \tag{5}$$

---

[87]In [68] there is $+17x^2$ in place of $-17x^2$ in the final expression.

Equation (5) is implicit in MacLaurin's demonstration of Proposition IV (see NPIV, equations (4) and (5), pp. 178–179). However, he gives no details for the derivation of (4) from (5); it does come out if we proceed as we did above in obtaining (3) from (1), but use the approximation $\frac{x}{d} \approx \frac{15V}{8A}$ in place of that used in the final step above.

Remark II (pp. 134–136). Let us first note the important relations

$$a^2 - e^2 = c^2, \qquad b^2 - e^2 = f^2, \qquad a^2 - b^2 = g^2, \tag{3}$$

where (see diagrams below) $a = |CA|$, $b = |CB|$, $c = |CF|$, $e = |CP|$, $f = |Cf|$ and $g = |Cg|$. The first of these comes from the fact that F is a focus of the ellipse $PApa$; likewise for the second we have that f is a focus of the ellipse $PBpb$ and for the third that g is a focus of the ellipse $ABab$.

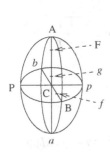

Diagram for
Proposition VIII
MacLaurin's Fig. 11

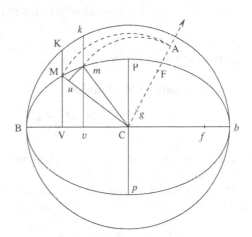

Figure from end of "Remarks"

Referring to the discussion in NLV (pp. 166–169) we note that

$$\int_{[A,C]} |RK|\, d|AR|$$

is no longer the same for each section. The expression for the attraction which is given in equation (3) of NLV (p. 168) now becomes

$$\frac{k}{|AC|} \int_0^{2\pi} \int_{[A,C]} |RK(\theta)|\, d|AR|\, d\theta,$$

where θ denotes the angle bCM, and by symmetry this is

$$\frac{4k}{a} \int_0^{\pi/2} \int_{[A,C]} |RK(\theta)|\, d|AR|\, d\theta.$$

Now we can evaluate $\int_{[A,C]} |RK(\theta)|\, d|AR|$ for any given θ by means of the procedure of Proposition II to get from equation (2) of NPII (p. 170) its value

$$\frac{2a^2|CM|^2}{(a^2 - |CM|^2)^{3/2}}\left(a\ln\sqrt{\frac{a + (a^2 - |CM|^2)^{1/2}}{a - (a^2 - |CM|^2)^{1/2}}} - \sqrt{a^2 - |CM|^2}\right)$$

$$= \frac{2a^2(a^2 - x^2)}{x^3}\left(a\ln\sqrt{\frac{a + x}{a - x}} - x\right),$$

and so the required attraction is

$$8ka\int_0^{\pi/2}\frac{(a^2 - x^2)}{x^3}\left(a\ln\sqrt{\frac{a + x}{a - x}} - x\right)d\theta.$$

If we compare this as usual with the attraction at A of the sphere on Aa as diameter,[88] we get (cf. equation (3) of NPII, p. 171)

$$\frac{|\text{Attraction at } A \text{ of figure}|}{|\text{Attraction at } A \text{ of sphere}|} = \frac{6}{\pi}\int_0^{\pi/2}\frac{(a^2 - x^2)}{x^3}\left(a\ln\sqrt{\frac{a + x}{a - x}} - x\right)d\theta.$$

$$(4)$$

This corresponds to MacLaurin's

$$\int\frac{3CM^2 \times \overline{L - x}}{x^3} \times \frac{mu}{CM}.$$

Note that, if we put $\delta\theta = \angle mCu$, then $\operatorname{arc} mu = |Cm| \times \delta\theta \approx |CM| \times \delta\theta$; in the limit $(\operatorname{arc} mu)/CM$ becomes $d\theta$.

We need to determine the relationship between x and θ $(0 \leq \theta \leq \pi/2)$. For this we have $a^2 - x^2 = |CM|^2$ and so

$$|CV| = \sqrt{a^2 - x^2}\cos\theta, \qquad |VM| = \sqrt{a^2 - x^2}\sin\theta;$$

the equation for the ellipse $PBpb$ then gives

$$\frac{a^2 - x^2}{b^2}\cos^2\theta + \frac{a^2 - x^2}{e^2}\sin^2\theta = 1,$$

from which we obtain, after some manipulation and use of equations (3),

$$\theta = \sin^{-1}\left(\frac{e}{f}\sqrt{\frac{x^2 - g^2}{a^2 - x^2}}\right), \quad\text{therefore}\quad \frac{d\theta}{dx} = \frac{ebx}{(a^2 - x^2)\sqrt{c^2 - x^2}\sqrt{x^2 - g^2}}.$$

Note also that the quantity $x = \sqrt{|CA|^2 - |CM|^2}$ takes on values between $\sqrt{a^2 - b^2} = g$ and $\sqrt{a^2 - e^2} = c$ as θ varies between 0 and $\pi/2$. Thus, changing the variable from θ to x in the integral in (4), we obtain for it

---

[88] As noted at the end of NLV (p. 169), $|\text{Attraction at } A \text{ of sphere}| = \frac{4}{3}k\pi a$.

$$\frac{6}{\pi}\int_g^c \frac{(a^2-x^2)}{x^3}\left(a\ln\sqrt{\frac{a+x}{a-x}}-x\right)\frac{ebx}{(a^2-x^2)\sqrt{c^2-x^2}\sqrt{x^2-g^2}}\,dx$$

$$=\frac{6eb}{\pi}\int_g^c \frac{1}{x^2\sqrt{c^2-x^2}\sqrt{x^2-g^2}}\left(a\ln\sqrt{\frac{a+x}{a-x}}-x\right)dx\,.$$

Now we expand the logarithm and integrate term-by-term; we need one term more than MacLaurin considers in his *Remarks* II in order to produce the expression given in Proposition VIII:

$$a\ln\sqrt{\frac{a+x}{a-x}}-x=\frac{1}{2}a\ln\frac{1+\frac{x}{a}}{1-\frac{x}{a}}-x=\frac{x^3}{3a^2}+\frac{x^5}{5a^4}+\frac{x^7}{7a^6}+\dots\,,$$

and the integral becomes

$$\frac{6eb}{\pi}\left(\frac{1}{3a^2}\int_g^c \frac{x}{\sqrt{c^2-x^2}\sqrt{x^2-g^2}}\,dx+\frac{1}{5a^4}\int_g^c \frac{x^3}{\sqrt{c^2-x^2}\sqrt{x^2-g^2}}\,dx\right.$$

$$\left.+\frac{1}{7a^6}\int_g^c \frac{x^5}{\sqrt{c^2-x^2}\sqrt{x^2-g^2}}\,dx+\dots\right)\,.$$

Following MacLaurin, we substitute $z=\sqrt{x^2-g^2}$ with $\frac{dz}{dx}=\frac{x}{\sqrt{x^2-g^2}}$ to get

$$\frac{|\text{Attraction at }A\text{ of figure}|}{|\text{Attraction at }A\text{ of sphere}|}$$

$$=\frac{6eb}{\pi}\left(\frac{1}{3a^2}\int_0^{\sqrt{c^2-g^2}}\frac{1}{\sqrt{c^2-g^2-z^2}}\,dz+\frac{1}{5a^4}\int_0^{\sqrt{c^2-g^2}}\frac{z^2+g^2}{\sqrt{c^2-g^2-z^2}}\,dz\right.$$

$$\left.+\frac{1}{7a^6}\int_0^{\sqrt{c^2-g^2}}\frac{(z^2+g^2)^2}{\sqrt{c^2-g^2-z^2}}\,dz+\dots\right)\,. \tag{5}$$

To evaluate these integrals we can use the standard formulae (for $\alpha\ne0$)

$$\int\frac{1}{\sqrt{\alpha^2-u^2}}\,du=\sin^{-1}\frac{u}{\alpha}+\text{constant}\,,$$

$$\int\sqrt{\alpha^2-u^2}\,du=\frac{u}{2}\sqrt{\alpha^2-u^2}+\frac{\alpha^2}{2}\int\frac{1}{\sqrt{\alpha^2-u^2}}\,du\,,$$

$$\int(\alpha^2-u^2)^{3/2}\,du=\frac{u}{4}(\alpha^2-u^2)^{3/2}+\frac{3\alpha^2}{4}\int\sqrt{\alpha^2-u^2}\,du\,.$$

Then the first integral in (5) is

$$\int_0^{\sqrt{c^2-g^2}} \frac{1}{\sqrt{c^2-g^2-z^2}}\, dz = \left[\sin^{-1}\frac{z}{\sqrt{c^2-g^2}}\right]_0^{\sqrt{c^2-g^2}} = \frac{\pi}{2}\,;$$

the second is

$$\int_0^{\sqrt{c^2-g^2}} \frac{z^2+g^2}{\sqrt{c^2-g^2-z^2}}\, dz$$

$$= \int_0^{\sqrt{c^2-g^2}} -\sqrt{c^2-g^2-z^2} + \frac{c^2}{\sqrt{c^2-g^2-z^2}}\, dz$$

$$= \left[-\frac{z}{2}\sqrt{c^2-g^2-z^2} + \frac{1}{2}(c^2+g^2)\sin^{-1}\frac{z}{\sqrt{c^2-g^2}}\right]_0^{\sqrt{c^2-g^2}}$$

$$= \frac{1}{2}(c^2+g^2)\frac{\pi}{2} = \frac{\pi}{4}\left(|CF|^2+|Cg|^2\right)\,;$$

and the third is

$$\int_0^{\sqrt{c^2-g^2}} \frac{(z^2+g^2)^2}{\sqrt{c^2-g^2-z^2}}\, dz = \int_0^{\sqrt{c^2-g^2}} \frac{\left((c^2-g^2-z^2)-c^2\right)^2}{\sqrt{c^2-g^2-z^2}}\, dz$$

$$= \int_0^{\sqrt{c^2-g^2}} (c^2-g^2-z^2)^{3/2} - 2c^2\sqrt{c^2-g^2-z^2} + \frac{c^4}{\sqrt{c^2-g^2-z^2}}\, dz$$

$$= \left[\frac{z}{4}(c^2-g^2-z^2)^{3/2} + \left(\frac{3}{4}(c^2-g^2)-2c^2\right)\times\frac{z}{2}\sqrt{c^2-g^2-z^2}\right.$$

$$\left. + \left(\left(\frac{3}{4}(c^2-g^2)-2c^2\right)\frac{(c^2-g^2)}{2}+c^4\right)\sin^{-1}\frac{z}{\sqrt{c^2-g^2}}\right]_0^{\sqrt{c^2-g^2}}$$

$$= \frac{\pi}{16}\left((-5c^2-3g^2)(c^2-g^2)+8c^4\right) = \frac{\pi}{16}(3c^4+2c^2g^2+3g^4)$$

$$= \frac{\pi}{16}(3|CF|^4+2|CF|^2|Cg|^2+3|Cg|^4)\,.$$

Substituting these values into (5), we find that

$$\frac{|\text{Attraction at } A \text{ of figure}|}{|\text{Attraction at } A \text{ of sphere}|}$$

$$= \frac{|CB|\times|CP|}{|CA|^2}\left(1+\frac{3|CF|^2+3|Cg|^2}{10|CA|^2}\right.$$

$$\left. +\frac{9|CF|^4+6|CF|^2|Cg|^2+9|Cg|^4}{56|CA|^4}+\dots\right),$$

precisely as MacLaurin asserts in Proposition VIII.

Remark III (p. 136). MacLaurin brings in another idea from Newton, namely, that tidal forces are inversely proportional to the *cubes* of the distances (see Propositions XXXVI and XXXVII of Book III of the *Principia* [85] ([15, 63]) and my Introduction, p. 91); the stated expressions for the forces come immediately from this. Then, from

$$\frac{\dfrac{Ld^3}{X^3} + \dfrac{SD^3}{Z^3}}{\dfrac{Ld^3}{x^3} + \dfrac{SD^3}{z^3}} = \frac{m}{n},$$

we obtain

$$\frac{nLd^3}{X^3} + \frac{nSD^3}{Z^3} = \frac{mLd^3}{x^3} + \frac{mSD^3}{z^3},$$

or equivalently

$$\frac{L}{S}\left(\frac{nd^3}{X^3} - \frac{md^3}{x^3}\right) = \frac{mD^3}{z^3} - \frac{nD^3}{Z^3};$$

MacLaurin's expression for the ratio L/S now follows.

Appendix III

III.1. Concerning Ellipses

Diameters of ellipses have an important role to play in MacLaurin's discussions. A *diameter* is a chord which passes through the centre, so the diameter from the point $(a\cos\theta, b\sin\theta)$ on the ellipse with equation $x^2/a^2 + y^2/b^2 = 1$ has equation

$$bx\sin\theta - ay\cos\theta = 0.$$

This diameter meets the ellipse again where

$$x = -a\cos\theta = a\cos(\theta + \pi), \quad y = -b\sin\theta = b\sin(\theta + \pi),$$

that is, at the point with parameter $\theta + \pi$. The gradient (slope) of the tangent at *both* extremities of the diameter is

$$\frac{b\cos\theta}{-a\sin\theta} = -\frac{b}{a}\cot\theta,$$

unless $\sin\theta = 0$, in which case the tangents are perpendicular to the x-axis.

The *conjugate diameter* for a given diameter may be defined as the diameter which is parallel to the tangents at the extremities of the given diameter. Thus in the present case the conjugate diameter has equation

$$bx\cos\theta + ay\sin\theta = 0, \quad \text{that is,} \quad bx\sin(\theta + \frac{\pi}{2}) - ay\cos(\theta + \frac{\pi}{2}) = 0;$$

its extremities are the points with parameters $\theta + \frac{\pi}{2}$ and $\theta + \frac{3\pi}{2}$ and its conjugate diameter has equation

$$bx\sin(\theta + \pi) - ay\cos(\theta + \pi) = 0, \quad \text{that is,} \quad bx\sin\theta - ay\cos\theta = 0.$$

Thus the conjugate of the conjugate is just the original diameter. We may therefore speak of a pair of conjugate diameters (see Fig. 1 below). A *semidiameter* is a line from the centre to a point on the ellipse; *conjugate semidiameters* occur when the corresponding whole diameters are conjugate. In the case of a circle, conjugate (semi)diameters are (semi)diameters which are at right angles to each other.

As a preparation for his study of the figure of the Earth in [69] MacLaurin develops various properties of ellipses in Articles 609–627. The method used is to deduce these by means of a projection method from corresponding, more easily proved, properties of circles.

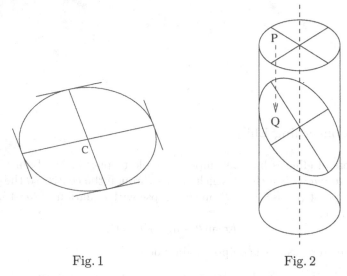

Fig. 1   Fig. 2

The section of a right circular cylinder by a plane which is not parallel to its axis is an ellipse (a circle if the plane is normal to the axis of the cylinder) (see Fig. 2); this of course requires proof, which MacLaurin gives in Article 610. Clearly, any ellipse can be generated in this way by taking the diameter of the circle equal to the length of the minor axis and choosing the appropriate inclination of the plane to produce the desired length of major axis. Having chosen a plane normal to the axis of the cylinder and a sectioning plane to produce an ellipse, we project points from the first plane into the second plane (and reciprocally) by means of lines parallel to the axis of the cylinder. Among the properties of this projection are the following, which are more or less obvious:

(i)   parallel lines project onto parallel lines (Article 611);
(ii) ratios of segments of the same line or of parallel lines are preserved on projection (Article 611);
(iii) a tangent to the circle projects onto a tangent to the ellipse (Article 612);
(iv) diameters of the circle which are at right angles to each other project onto conjugate diameters of the ellipse (see above) (Article 612).

As a simple illustration of this method we prove that a diameter of an ellipse bisects all chords parallel to its conjugate – these chords are called *ordinates* to the diameter. But this is obvious in a circle – a diameter bisects any chord at right angles to it – and the result for the ellipse then follows from (iv), (i) and (ii).

In his essay MacLaurin quotes several properties of ellipses which can be established easily in this way (see NLI pp. 141–143, NPII p. 170, NPIII p. 174). Some further discussion will be found in Section 3 of [106].

III.2. Sections of Spheroids

By a *spheroid* MacLaurin means the solid formed by rotating an ellipse about one of its axes; its *equatorial plane* is the plane through the centre which is perpendicular to the axis of rotation. He applies to great effect the following two properties of spheroids; these are stated without proof in his essay but proofs are given in Article 633 of [69].

(i) If a plane meets two similar, similarly situated spheroids,[89] then the sections are similar ellipses.

(ii) If a plane perpendicular to the equatorial plane of a spheroid meets the spheroid, then the section is similar to the generating ellipse.

Because of the importance of these results to MacLaurin's arguments it seems desirable to establish them here. The analytic proofs below may be more accessible to the modern reader than those given by MacLaurin.

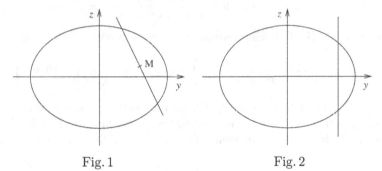

Fig. 1   Fig. 2

We may assume for the spheroid an equation of the form

$$\frac{x^2}{a^2} + \frac{y^2}{a^2} + \frac{z^2}{b^2} = 1 .$$

This represents the spheroid generated by rotating the ellipse in the z, y-plane with equation $y^2/a^2 + z^2/b^2 = 1$ about the z-axis; here we may have $a < b$, $a = b$ or $a > b$ $(a, b > 0)$. Clearly, any plane which cuts the spheroid and is parallel to the x, y-plane does so in a circular section, so (i) is trivial in this case. We now consider planes which are not parallel to the x, y-plane. Because of the symmetry of the spheroid it is enough to consider planes which are

---

[89]That is to say, they are generated by rotating about the same axis two ellipses which have the same centre, corresponding axes along the same line, and the same ratio of major axis to minor axis.

perpendicular to the y, z-plane, since any other situation can be brought to this by rotation. Such a plane has an equation of the form

$$y = kz + c \tag{1}$$

for suitable constants k and c. Now (Fig. 1) the straight line in the y, z-plane with the same equation meets the generating ellipse where

$$\frac{(kz + c)^2}{a^2} + \frac{z^2}{b^2} = 1,$$

that is, where

$$\left(\frac{k^2}{a^2} + \frac{1}{b^2}\right) z^2 + \frac{2kc}{a^2} z + \frac{c^2}{a^2} - 1 = 0.$$

The sum of the roots is

$$-\frac{2kc}{a^2} \left(\frac{k^2}{a^2} + \frac{1}{b^2}\right)^{-1} = \frac{-2kcb^2}{k^2b^2 + a^2},$$

and so M, the midpoint of the line segment joining the points of intersection, is given by

$$z = \frac{-kcb^2}{k^2b^2 + a^2}, \quad y = \frac{-k^2cb^2}{k^2b^2 + a^2} + c = \frac{a^2c}{k^2b^2 + a^2}.$$

We now use the following parametric representation of the plane:

$$\mathbf{r} = \left(0, \frac{a^2c}{k^2b^2 + a^2}, \frac{-kcb^2}{k^2b^2 + a^2}\right) + \xi(1, 0, 0) + \frac{\eta}{\sqrt{k^2 + 1}}(0, k, 1).$$

Note that $(1, 0, 0)$ and $(k^2 + 1)^{-1/2}(0, k, 1)$ are orthogonal unit vectors which are orthogonal to $(0, 1, -k)$, a normal to the plane, and the first triple is just the point M in the x, y, z-system. The plane therefore meets the spheroid where

$$\frac{\xi^2}{a^2} + \frac{1}{a^2}\left(\frac{a^2c}{k^2b^2 + a^2} + \frac{k\eta}{\sqrt{k^2 + 1}}\right)^2 + \frac{1}{b^2}\left(\frac{-kcb^2}{k^2b^2 + a^2} + \frac{\eta}{\sqrt{k^2 + 1}}\right)^2 = 1,$$

that is,

$$\frac{\xi^2}{a^2} + \left(\frac{k^2}{a^2(k^2 + 1)} + \frac{1}{b^2(k^2 + 1)}\right)\eta^2 = 1 - \frac{a^2c^2}{(k^2b^2 + a^2)^2} - \frac{k^2c^2b^2}{(k^2b^2 + a^2)^2}$$

$$= 1 - \frac{c^2}{k^2b^2 + a^2},$$

or,

$$\frac{\xi^2}{a^2\left(1 - \frac{c^2}{k^2b^2 + a^2}\right)} + \frac{\eta^2}{\frac{a^2b^2(k^2 + 1)}{k^2b^2 + a^2}\left(1 - \frac{c^2}{k^2b^2 + a^2}\right)} = 1. \tag{2}$$

Thus, provided

$$\frac{c^2}{k^2 b^2 + a^2} < 1,$$

the section of the spheroid by the plane is an ellipse with centre M and axes parallel to $(1,0,0)$ and $(0,k,1)$; moreover, the ratio of its ξ-axis to its η-axis is

$$\sqrt{\frac{k^2 b^2 + a^2}{b^2 (k^2 + 1)}} = \frac{1}{\sqrt{k^2 + 1}} \left(k^2 + \frac{a^2}{b^2} \right)^{1/2}.$$

Since this depends only on the ratio $\frac{a}{b}$ for given k, it follows that, if the plane meets any other similar, similarly situated spheroid, the two sections will be similar ellipses. Note that, if

$$\frac{c^2}{k^2 b^2 + a^2} = 1,$$

the plane is tangential to the spheroid and, if it is greater than 1, there is no intersection.

Suppose now for (ii) that the plane is normal to the equatorial plane (Fig. 2), so that (1) becomes $y = c$ and $k = 0$. Equation (2) reduces to

$$\frac{\xi^2}{a^2 \left(1 - \frac{c^2}{a^2} \right)} + \frac{\eta^2}{b^2 \left(1 - \frac{c^2}{a^2} \right)} = 1.$$

The ratio of the axes is now a/b, so we have an ellipse which is similar to the generating ellipse (provided $c^2/a^2 < 1$).

III.3. Attraction

Newton's law of gravitation asserts that the force of attraction between two particles of masses m_1 and m_2 and at distance r apart is given by

$$\frac{G m_1 m_2}{r^2},$$

where G is a constant, and it acts along the line joining the particles. It follows from this that the attraction at a point $P(x_0, y_0, z_0)$ from a uniform body V is proportional to

$$\int\int\int_V \frac{1}{(x - x_0)^2 + (y - y_0)^2 + (z - z_0)^2} \hat{\mathbf{r}} \, dx \, dy \, dz, \tag{1}$$

provided the integral converges; here $\hat{\mathbf{r}}$ is the unit vector in the direction from P to the varying point (x, y, z), that is to say,

$$\hat{\mathbf{r}} = \frac{(x - x_0, y - y_0, z - z_0)}{\sqrt{(x - x_0)^2 + (y - y_0)^2 + (z - z_0)^2}}.$$

We may therefore write (1) as

$$\int\int_V\int \frac{(x-x_0, y-y_0, z-z_0)}{((x-x_0)^2+(y-y_0)^2+(z-z_0)^2)^{3/2}}\, dx\, dy\, dz\,. \qquad (2)$$

If we make the change of variables

$$x = x_0 + \rho\sin\phi\cos\theta\,, \quad y = y_0 + \rho\sin\phi\sin\theta\,, \quad z = z_0 + \rho\cos\phi$$

(spherical polar coordinates with origin (x_0, y_0, z_0)), we obtain

$$\int\int_{V'}\int \sin\phi\,(\sin\phi\cos\theta, \sin\phi\sin\theta, \cos\phi)\, d\rho\, d\phi\, d\theta\,, \qquad (3)$$

where V' is the corresponding region in the (ρ, ϕ, θ)-system. We note for application (see NL3, p. 149) that the integrand $\mathbf{h}(\phi, \theta)$ in (3) is independent of ρ and it satisfies the identity

$$\mathbf{h}(\pi - \phi, \theta + \pi) = -\mathbf{h}(\phi, \theta)\,. \qquad (4)$$

If $\mathbf{u} = (u_1, u_2, u_3)$ is a *unit* vector, the component of (2) in the direction of $\mathbf{u}$ is

$$\int\int_V\int \frac{u_1(x-x_0) + u_2(y-y_0) + u_3(z-z_0)}{((x-x_0)^2+(y-y_0)^2+(z-z_0)^2)^{3/2}}\, dx\, dy\, dz\,; \qquad (5)$$

the equivalent version from (3) is

$$\int\int_{V'}\int u_1\sin^2\phi\cos\theta + u_2\sin^2\phi\sin\theta + u_3\sin\phi\cos\phi\, d\rho\, d\phi\, d\theta\,. \qquad (6)$$

Several of MacLaurin's calculations involve these equations implicitly (see, for example, NLIII pp. 146–147, 149, NLIV pp. 154–156, NLV pp. 168–169, NPIII p. 176).

III.4. Extract from *"Reflexions sur les Observations des Marées"* (1713) by J. Cassini [22]

(pp. 286–288)

Nous avons remarqué dans les Memoires précedents que les diverses distances de la Lune à la Terre causent une trés grande varieté dans la hauteur de Marées. Cela se confirme par ces dernieres Observations, car le 28. Decembre 1712. jour de la Pleine Lune, la distance de cette Planette à la Terre, étant de 936. parties dont le rayon est 1000, c'est-à-dire, la Lune étant fort prés de son Perigée, on observa le 30. Decembre au matin, jour de la plus grande Marée, la hauteur de la Pleine Mer de 19. pieds 2. pouces au-dessus du point fixe, et celle de la Basse Mer de 1. pied 8. pouces au-dessous de ce point, de sorte que la Mer avoit monté ce jour-là de la hauteur de 20. pieds 10. pouces.

...

Il faut remarquer que dans la Nouvelle Lune Perigée du 28. Decembre 1712. sa déclinaison étoit de 23^d $0'$ Meridionale, fort éloignée de l'Equinoct-ial, & par consequent sa pression sur la Terre devoit être moins grande que lorsque la Lune étant à peu prés à égale distance de la Terre, elle se trouve en même temps plus prés de l'Equateur.

En effet nous trouvons que le 24. Fevrier 1713. jour de la Nouvelle Lune, sa distance à la Terre étant de 953. c'est-à-dire, prés de son Perigée, & sa déclinaison de 5^d Meridionale prés de l'Equateur, la hauteur de la Pleine Mer fut observée le 26. Fevrier au matin de 21. pieds 2. pouces, qui est la plus Haute Marée que l'on ait observé à Brest dans l'espace de prés de deux années. La Basse Mer suivante fut observée de 1. pied 3. pouces au-dessous du point fixe, de sorte que la Mer monta ce jour-là de la hauteur de 22. pieds 5. pouces.

Le 12. Mars suivant, jour de la Pleine Lune, sa distance à la Terre étant de 1032. assés prés de son Apogée & sa déclinaison Meridionale d'un degré, c'est-à-dire, prés de l'Equateur, on observa le 13. Mars suivant, jour de la plus grande Marée, la hauteur de la Pleine Mer de 18. pieds 2. pouces, & celle de la Basse Mer de 0. pied 0. pouce, de sorte que l'élevation de la Mer n'a été ce jour-là que de 18. pieds 2. pouces, moindre de 4. pieds trois pouces que dans l'Observation précedente où la Lune étoit prés de son Perigée, mais plus grande de 1. pied 9. pouces que dans l'Observation du 11. Janvier 1713. raportée ci-devant, où la Lune étant prés de son Apogée, sa déclinaison Septemtrionale étoit de 20. degrés.

(pp. 289–290)

À l'égard de la distance du Soleil à la Terre, comme elle est plus petite vers le Solstice d'Hyver où le Soleil est presentement prés de son Perigée, qu'au Solstice d'Eté où il est prés de son Apogée, les Marées doivent être plus grandes en Hyver qu'en Eté, toutes choses égales, comme on l'observe en effet. Car le 30. Juin[90] 1711. jour de la Pleine Lune, la distance de la Lune à la Terre étant de 960. & sa déclinaison de 25^d $29'$; le Soleil étant aussi dans son Apogée, on observa le premier Juillet au soir la hauteur de la plus grande Marée de 17. pieds 10. pouces. Le 8. Janvier suivant, jour de la Nouvelle Lune, la distance de la Lune à la Terre étant de 951. & sa déclinaison de 23^d $0'$ à peu prés de même que le 30. Juin; le Soleil étant alors prés de son Perigée, on observa le 10. Janvier au matin, la hauteur de la plus grande Marée de 19. pieds 10. pouces plus haute de 2. pieds que dans l'Observation précedente, où le Soleil étoit dans son Apogée. Le 19. Juin suivant la distance de la Lune à la Terre étant de 936. & sa déclinaison Meridionale de 24^d $50'$,

---

[90]The version in [22] has "Juillet" here, which does not seem right as the next date mentioned is 1 July.

le Soleil étant alors prés de son Apogée, la hauteur de la plus grande Marée
fut observé le 21. Juin au soir de 18. pieds 4. pouces plus petite d'un pied
six pouces que dans l'Observation précedente. Enfin le 28. Decembre 1712.
le Soleil étant dans son Perigée, la distance de la Lune à la Terre étant de
936. & sa déclinaison Meridionale de 23. degrés, la hauteur de la plus grande
Marée fut observée le 30. Decembre de 19. pieds 2. pouces, plus grande de
10. pouces que le 19. Juin où le Soleil étoit prés de son Apogée, & la Lune à
peu prés à égale distance de la Terre.

(Translation)

(pp. 286–288)

We have noted in the preceding Memoirs that the different distances from
the Moon to the Earth bring about a very great variation in the height of
the tides. That is confirmed by these recent observations, for on 28 December
1712, the day of the Full Moon, the distance from this planet to the Earth
being 936 parts of which the radius is 1000, that is to say, the Moon being
very near to its perigee, it was observed on the morning of 30 December, the
day of the greatest tide, that the height of the high tide was 19 pieds 2 pouces
above the fixed point, and that of the low water was 1 pied 8 pouces below
this point, so that the sea had risen by a height of 20 pieds 10 pouces on that
day.

. . .

It must be noted that in the New Moon perigee of 28 December 1712 its
declination was 23° 0′ of the meridian, far removed from the equator, and
in consequence its pressure on the Earth must be less great than when the
Moon, being at approximately equal distance from the Earth, is at the same
time nearer to the equator.

In fact, we find that on 24 February 1713, the day of the New Moon,
its distance to the Earth being 953, that is to say, near its perigee, and its
declination 5° of the meridian near the equator, the height of the high tide
was observed on the morning of 26 February at 21 pieds 2 pouces, which is
the highest tide which has been observed at Brest in the space of nearly two
years. The low water following was observed at 1 pied 3 pouces below the
fixed point, so that the sea rose by a height of 22 pieds 5 pouces on that day.

On 12 March following, the day of the Full Moon, its distance from the
Earth being 1032, quite close to its apogee, and its meridional declination
one degree, that is to say, near the equator, it was observed on 13 March
following, the day of the greatest tide, that the height of the high tide was
18 pieds 2 pouces, and that of the low water 0 pied 0 pouce, so that the rise of
the sea on that day was only 18 pieds 2 pouces, less by 4 pieds three pouces
than in the preceding observation where the Moon was near its perigee, but
larger by 1 pied 9 pouces than in the observation of 11 January 1713 reported
previously, where its northern declination was 20 degrees, the Moon being
near its apogee.

(pp. 289–290)

Concerning the distance from the Sun to the Earth, since it is smaller towards the winter solstice, where the Sun is at present near its perigee, than at the summer solstice, where it is near its apogee, the tides must be larger in winter than in summer, other things being equal, as is indeed observed. For on 30 June[91] 1711, the day of the Full Moon, the distance from the Moon to the Earth being 960 and its declination 25° 29′ and the Sun being also in its apogee, on the evening of the first of July the height of the greatest tide was observed at 17 pieds 10 pouces. On 8 January following, the day of the New Moon, the distance from the Moon to the Earth being 951 and its declination 23° 0′, approximately the same as on 30 June and the Sun being then near its perigee, on the morning of 10 January it was observed that the height of the largest tide was 19 pieds 10 pouces, higher by 2 pieds than in the previous observation, where the Sun was in its apogee. On 19 June following, the distance from the Moon to the Earth being 936 and its meridional declination 24° 50′, the Sun being then near its apogee, the height of the greatest tide was observed on the evening of 21 June at 18 pieds 4 pouces, smaller by one pied six pouces than in the preceding observation. Finally, on 28 December 1712, the Sun being in its perigee, the distance from the Moon to the Earth being 936 and its meridional declination 23 degrees, the height of the greatest tide was observed on 30 December at 19 pieds 2 pouces, greater by 10 pouces than on 19 June when the Sun was near its apogee and the Moon at approximately equal distance from the Earth.

III.5. Proof of Corollary 4 of Lemma I from [86]

The following Theorem and its proof were added by the editors as a footnote to MacLaurin's Lemma I in the version of his essay in [86]. I have redrawn the original diagrams, which are not very accurate, and have added some explanatory notes at the end, which are referenced by Greek letters inserted in the text.

This Lemma is proposed for demonstrating the 4th Corollary; this Corollary is reduced to the following Proposition, which can be demonstrated very easily by analysis.

THEOREM.

From any point on an ellipse let three lines PH, PM, Pm be drawn to the ellipse, the first of which PH is parallel to the axis, while the others PM, Pm make any equal angles MPH, mPH (α) with it; from the points P, H, M and m let PD, Hd, QMR, mqr be drawn perpendicular to PH and to

---

[91]The version in [22] has "Juillet" (July) here, which does not seem right as the next date mentioned is 1 July.

the axis and on Dd let an ellipse similar to the first be described, and from
the point D let lines DN, Dn parallel to the lines Pm, PM be drawn to
the ellipse, and finally let Nn be drawn which cuts the axis in V; I say that
$2DV = PQ + Pq = DR + Dr$, if the points Q and q lie on the same side of
the point P, or that $2DV = PQ - Pq = DR - Dr$, if the points Q and q lie
on opposite sides of the point P.

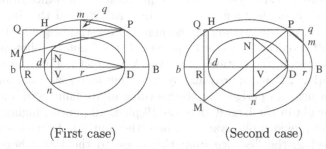

(First case)   (Second case)

First, since by construction the lines DN, Dn make equal angles with the
axis Dd, it is easily deduced that the line NVn is perpendicular to the axis
and so, if the radius is to the tangent of angle QPM as 1 to t, and DV is
called z, there will be $NV = tz$; and likewise, if PQ or Pq or their equals
DR or Dr are called x, then MQ or mq will be given by tx.

Let the major axis be to the minor axis in both ellipses as a to b and let
$BD = f$, $Db = g$, $DP = h$ and $Dd = g - f = l$; by the nature of the ellipse
(β) $a^2 : b^2 = fg : h^2$, and likewise there will be $a^2 : b^2 = z \times \overline{l - z} : t^2 z^2 =$
$l - z : t^2 z$, hence $a^2 : \frac{b^2}{t^2} = l - z : z$ and by composition (γ)

$$\frac{t^2 a^2 + b^2}{t^2} : \frac{b^2}{t^2} = a^2 t^2 + b^2 : b^2 = l : z \text{ and so } \frac{b^2 l}{a^2 t^2 + b^2} = DV .$$

Moreover, in the first case, in which Q and q are on the same side of the
point P, there will be (δ) $RM = h - tx$, or $tx - h$, and $rm = h + tx$, and BR
or Br will be $f + x$, and Rb or rb will be $g - x$; hence, from the nature of the
ellipse there will be (β)

$$a^2 : b^2 = \overline{f + x} \times \overline{g - x} : \overline{h \mp tx}^2 = fg + gx - fx - x^2 : h^2 \mp 2htx + t^2 x^2$$
$$= lx - x^2 : \mp 2htx + t^2 x^2$$

(the terms $fg : h^2$, which are in the same ratio, having been taken away
from both terms, respectively (ϵ), and l having been put in place of $g - f$)
$= l - x : \mp 2ht + t^2 x$, and hence is obtained $a^2 t^2 x \mp 2a^2 ht = b^2 l - b^2 x$

and, following transposition and reduction of the terms,

$$x = \frac{b^2 l \pm 2a^2 ht}{a^2 t^2 + b^2} .$$

Consequently, if the sum of the two lines DR, Dr, which are given by the
individual values of x, is taken, there will be

$$DR + Dr = PQ + Pq = \frac{2b^2l}{a^2t^2 + b^2},$$

twice the value found previously for DV.

But in the other case, in which Q and q lie on opposite sides of the point P, there will be (ζ) $RM = tx - h$, and $rm = h - tx$, and there will be $BR = f + x$ and $Br = f - x$, $Rb = g - x$ and $rb = g + x$. Hence, from the nature of the ellipse there will be (β)

$$a^2 : b^2 = \overline{f \pm x} \times \overline{g \mp x} : h^2 - 2htx + t^2x^2$$

$$= fg \pm gx \mp fx - x^2 : h^2 - 2htx + t^2x^2 = \pm lx - x^2 : -2htx + t^2x^2$$

(the terms $fg : h^2$ having been taken away (ϵ) and l having been used in place of $g - f$) $= \pm l - x : -2ht + t^2x$, and hence is obtained

$$a^2t^2x - 2hta^2 = \pm b^2l - b^2x$$

and, following transposition and reduction of the terms,

$$x = \frac{\pm b^2l + 2hta^2}{a^2t^2 + b^2}.$$

Consequently, if the difference of the two lines DR, Dr, which are given by the individual values of x, is taken, there will be (η)

$$DR - Dr = PQ - Pq = \frac{2b^2l}{a^2t^2 + b^2},$$

twice the value found previously for DV; therefore, $2DV = PQ \mp Pq$ according as Q and q are on the same or opposite sides of the point P (θ). Q.E.D.

Notes

(α). The angles should be described as MPQ and mPq to cover both cases.

(β). From the canonical equation for an ellipse, $\dfrac{\xi^2}{a^2} + \dfrac{\eta^2}{b^2} = 1$, we obtain

$$\frac{\eta^2}{b^2} = \frac{a^2 - \xi^2}{a^2} \quad \text{and so} \quad \frac{(a - \xi)(a + \xi)}{\eta^2} = \frac{a^2}{b^2} \quad (\eta \neq 0).$$

This is the version of the equation which is used in the proof; $a - \xi$ and $a + \xi$ represent the distances of a point on the major axis from its two extremities.

(γ). The ratio operation *componendo* is being used:

$$\frac{S}{T} = \frac{U}{V} \Rightarrow \frac{S + T}{T} = \frac{U + V}{V}.$$

(δ). The expressions for RM, BR, Rb have $x = DR$, while $x = Dr$ in those for rm, Br, rb.

(ϵ). Here we are using another ratio operation:

$$\frac{S}{T} = \frac{U}{V} \Rightarrow \frac{S-U}{T-V} = \frac{S}{T} \quad (T \neq V).$$

(ζ). The expressions for RM, BR have $x = DR$, while $x = Dr$ in those for rm, Br. In fact, $RM = \pm(tx - h)$ and $rm = \pm(h - tx)$, but this does not affect the subsequent argument since we use only their squares.

(η). Strictly, it is the modulus of the difference in general.

(θ). If we read $\mp$ as "minus or plus," then "same" and "opposite" should be interchanged.

III.6. MacLaurin's diagrams from [2] (courtesy of Glasgow University Library, reproduced with permission)

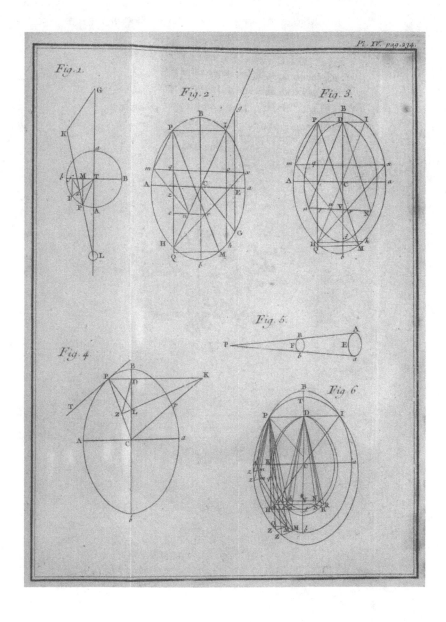

(III.6. MacLaurin's diagrams from [2], ctd.)

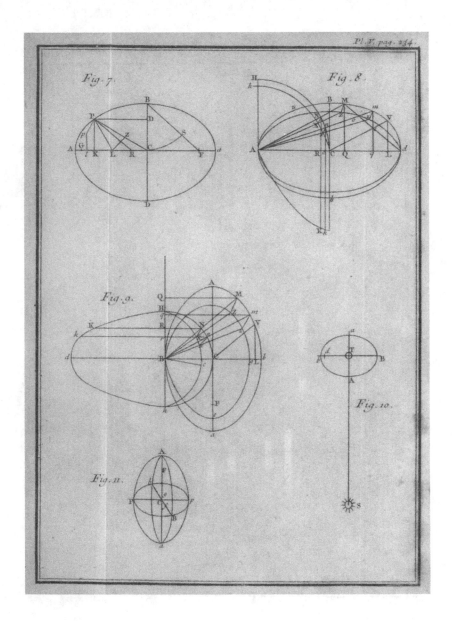

References

1. Académie des Sciences: Recueil des pièces qui ont remporté les prix. Paris, 1728.
2. Académie des Sciences: Recueil des pièces qui ont remporté les prix. Paris, 1750–1777.
3. E. J. Aiton: Galileo's theory of the tides. Ann. of Sci. **10**, 44–57 (1954).
4. E. J. Aiton: The contributions of Newton, Bernoulli and Euler to the theory of the tides. Ann. of Sci. **11**, 206–223 (1955–56).
5. E. J. Aiton: Descartes's theory of the tides. Ann. of Sci. **11**, 337–348 (1955–56).
6. E. J. Aiton: The vortex theory of planetary motions – I. Ann. of Sci. **13**, 249–264 (1957).
7. E. J. Aiton: The vortex theory of planetary motions – II. Ann. of Sci. **14**, 132–147 (1958).
8. E. J. Aiton: The vortex theory of planetary motions – III. Ann. of Sci. **14**, 157–172 (1958).
9. E. J. Aiton: The Cartesian theory of gravity. Ann. of Sci. **15**, 27–49 (1959).
10. E. J. Aiton: The Vortex Theory of Planetary Motions. Macdonald (London) and American Elsevier (New York), 1972.
11. H. G. Alexander: The Leibniz–Clarke Correspondence. Manchester University Press, Manchester and New York, 1956.
12. Daniel Bernoulli: Traité sur le Flux et Reflux de la Mer. [2] (1740), 53–191; reproduced in [86], Vol. III, 133–246.
13. Jean Bernoulli: Discours sur les Loix de la Communication du Mouvement. [1], (1727), 109 pp.
14. William Braikenridge: Exercitatio Geometrica de Descriptione Linearum Curvarum. London, 1733.
15. F. Cajori: Newton's Principia: Motte's Translation Revised. University of California Press, Berkeley, 1947.
16. D. S. L. Cardwell: Some factors in the early development of the concepts of power, work and energy. British J. Hist. Sci. **3**, 209–224 (1967).
17. Jacques Cassini: Reflexions sur les Observations du Flux et du Reflux de la Mer, faites à Dunquerque par M. Baert Professeur d'Hydrographie, pendant les années 1701 et 1702. Mém. de Math. et de Phys. in Histoire de l'Academie Royale des Sciences (Paris, 1710 (1712)), 318–341.
18. Jacques Cassini: Reflexions sur les Observations du Flux et du Reflux de la Mer, faites au Havre de Grace par M. Boissaye du Bocage Professeur d'Hydrographie, pendant les années 1701 et 1702. Mém. de Math. et de Phys. in Histoire de l'Academie Royale des Sciences (Paris, 1710 (1712)), 366–379.
19. Jacques Cassini: Reflexions sur les Observations des Marées faites à Brest et à Bayonne. Mém. de Math. et de Phys. in Histoire de l'Academie Royale des Sciences (Paris, 1710 (1712)), 380–385.

20. Jacques Cassini: Du Flux et Reflux de la Mer. Mém. de Math. et de Phys. in Histoire de l'Academie Royale des Sciences (Paris, 1712 (1714)), 86–96.

21. Jacques Cassini: Reflexions sur des nouvelles Observations du Flux et du Reflux de la Mer, faites au Port de Brest dans l'année 1712. Mém. de Math. et de Phys. in Histoire de l'Academie Royale des Sciences (Paris, 1713 (1716)), 14–30.

22. Jacques Cassini: Reflexions sur les Observations des Marées. Mém. de Math. et de Phys. in Histoire de l'Academie Royale des Sciences (Paris, 1713 (1716)), 267–290.

23. P. Antoine Cavalleri: Dissertation sur la cause physique du Flux et Reflux de la Mer. [2] (1740), 1–51.

24. S. Chandrasekhar: Ellipsoidal Figures of Equilibrium. Dover Publications, New York, 1987.

25. Alexis-Claude Clairaut: Investigationes aliquot, ex quibus probetur Terrae figuram secundum Leges attractionis in ratione inversa quadrati distantiarum maxime ad Ellipsin accedere debere. Phil. Trans. 40 (No. 445 Jan.–June 1737), 19–25 (1741).

26. Alexis-Claude Clairaut: An Inquiry concerning the Figure of such Planets as revolve about an Axis, supposing the Density continually to vary, from the Centre towards the Surface. Phil. Trans. 40 (No. 449 Aug.–Sept. 1738), 277–306 (1741).

27. Alexis-Claude Clairaut: Théorie de la Figure de la Terre. Paris, 1743.

28. Samuel Clarke: A Demonstration of the Being and Attributes of God. London, 1705 (reproduced in [30], pp. 513–577).

29. Samuel Clarke: A Letter from the Rev. Dr. Samuel Clarke to Mr. Benjamin Hoadly, F.R.S. occasion'd by the present Controversy among Mathematicians, concerning the Proportion of Velocity and Force in Bodies in Motion. Phil. Trans. 35 (No. 401 January, February March 1728), 381–388 (1727–28) (reproduced in [31], pp. 737–740.)

30. Samuel Clarke: The Works 1738, Vol. II. Garland Publishing, New York and London, 1978.

31. Samuel Clarke: The Works 1738, Vol. IV. Garland Publishing, New York and London, 1978.

32. I. B. Cohen: Isaac Newton's Theory of the Moon's Motion (1702). Dawson, Folkestone, 1975.

33. A. D. D. Craik: James Ivory, F.R.S., Mathematician: 'The most unlucky person that ever existed'. Notes Rec. R. Soc. Lond. 54, 223–247 (2000).

34. A. D. D. Craik: James Ivory's last papers on the 'Figure of the Earth' (with biographical additions). Notes Rec. R. Soc. Lond. 56, 187–204 (2002).

35. John Theophilus Desaguliers: A Course of Experimental Philosopy, Vol. II. London, 1744.

36. R. L. Emerson: The Philosophical Society of Edinburgh, 1737–1747. British J. Hist. Sci. 12, 154–191 (1979).

37. R. L. Emerson: Professors, Patronage and Politics: The Aberdeen Universities in the Eighteenth Century. Aberdeen University Press, Aberdeen, 1992.

38. Leonhard Euler: Inquisitio physica in causam fluxus ac refluxus maris. [2] (1740), 235–350; reproduced in [86], Vol. III, pp. 283–374 and in [40], pp. 19–124.

39. Leonhard Euler: Opera Omnia (second series) Vol. 12. Lausanne, 1954.

40. Leonhard Euler: Opera Omnia (second series) Vol. 31. Birkhäuser, Basel, 1996.

41. G. A. Gibson: Sketch of the history of mathematics in Scotland to the end of the 18th century. Proc. Edinburgh Math. Soc. Series 2, 1, 1–18, 71–93 (1927).

42. J. V. Grabiner: Was Newton's calculus a dead end? The continental influence of Maclaurin's Treatise of Fluxions. Amer. Math. Monthly **104**, 393–410 (1997).

43. J. V. Grabiner: Maclaurin and Newton: The Newtonian Style and the Authority of Mathematics. Science and Medicine in the Scottish Enlightenment (editors: C. W. J. Withers and P. Wood) (Tuckwell Press, East Linton, 2002), 143–171.

44. Willem 'sGravesande: Essai d'une Nouvelle Théorie du Choc des Corps. Journal littéraire de la Haye **XII**, 1–54 (1722) (also [49], Vol. 1, Part 1, 217–247).

45. Willem 'sGravesande: Supplément à l'Essai sur le Choc des Corps. Journal littéraire de la Haye **XII**, 190– (1722) (also [49], Vol. 1, Part 1, 247–251).

46. Willem 'sGravesande: Remarques sur la force des Corps en mouvement, et sur le Choc; précédées de quelques Réflexions sur la manière d'écrire de Mr. le Docteur Samuel Clarke. Journal littéraire de la Haye **XIII**, 189– , 407– (1730) (also [49], Vol. 1, Part 1, 251–268).

47. Willem 'sGravesande: Remarques touchant le Mouvement perpétuel. [49], Vol. 1, Part 1, 305–312.

48. Willem 'sGravesande: Elemens de Physique demontrez mathematiquement, et confirmez par des experiences. Leiden, 1746.

49. Willem 'sGravesande: Oeuvres Philosophiques et Mathématiques (editor: J. N. S. Allamand). Amsterdam, 1774.

50. J. L. Greenberg: The Problem of the Earth's Shape from Newton to Clairaut: The rise of mathematical science in eighteenth-century Paris and the fall of "normal" science. Cambridge, 1995.

51. David Gregory: Astronomiae Physicae et Geometricae Elementa. Oxford, 1702.

52. A. R. Hall: Mechanics and the Royal Society, 1668–70. British J. Hist. Sci. **3**, 24–38 (1966).

53. A. R. Hall and M. B. Hall: The Correspondence of Henry Oldenburg, Vol. 5 (1668–1669). University of Wisconsin Press, Madison, Milwaukee, and London, 1968.

54. B. Harrison (editor): Oxford Dictionary of National Biography. Oxford, 2004.

55. K. Holliday: Introductory Astronomy. Wiley, Chichester, 1999.

56. R. H. Hurlbutt: Hume, Newton, and the Design Argument. University of Nebraska Press, Lincoln, 1965.

57. Christiaan Huygens: De motu corporum ex percussione (1656). [60], 368–398; see also [61], Vol. XVI, 29–91, which includes a translation into French.

58. Christiaan Huygens: Extrait d'une lettre de M. Hugens à l'Auteur du Journal: Regles du mouvement dans le rencontre des Corps. Journal des Sçavans, 19–23 (1669).

59. Christiaan Huygens: A Summary Account of the Laws of Motion, communicated by Mr. Christian Hugens in a Letter to the R. Society, and first printed in French in the Journal des Scavans of March 18, 1669 st. n. Phil. Trans. IV No. 46 (April 12 1669), 925–928 (1669).

60. Christiaan Huygens: Opuscula posthuma. Leiden, 1703.

61. Christiaan Huygens: Oeuvres complètes de Christiaan Huygens. Société hollandaise de Sciences, la Haye, 1885–1950.

62. C. Innes and J. Robertson (editors): Munimenta alme Universitatis Glasguensis. Maitland Club, Glasgow, 1854.

63. A. Koyré and I. B. Cohen (editors): Isaac Newton's Philosophiae Naturalis Principia Mathematica (Third Edition). Cambridge, 1972.

64. J. W. Lubbock: Account of the "Traité sur le Flux et Réflux de la Mer," of Daniel Bernoulli; and a Treatise on the Attraction of Ellipsoids. London, 1830.

65. Colin MacLaurin: Geometria organica: sive descriptio linearum curvarum universalis. London, 1720.
66. Colin MacLaurin: Démonstration des Loix du Choc des Corps. [1] (1724), 24 pp.
67. Colin MacLaurin: A letter from Mr. Colin MacLaurin ... concerning the Description of Curve Lines. Phil. Trans. **39**, 143–165 (1735–36).
68. Colin MacLaurin: De Causa Physica Fluxus et Refluxus Maris. [2] (1740), 193–234; reproduced in [86], Vol. III, 247–282.
69. Colin MacLaurin: Treatise of Fluxions. Edinburgh, 1742.
70. Colin MacLaurin: An Account of Sir Isaac Newton's Philosophical Discoveries (editor: Patrick Murdoch). London, 1748.
71. Colin MacLaurin: Of the cause of the variation of the obliquity of the ecliptic. Essays and Observations, Physical and Literary **1**, 173–183 (1754).
72. Colin MacLaurin: Concerning the sudden and surprising changes observed in the surface of Jupiter's body. Essays and Observations, Physical and Literary **1**, 184–188 (1754).
73. Colin MacLaurin: An Account of Sir Isaac Newton's Philosophical Discoveries (editor: Patrick Murdoch). Georg Olms Verlag, Hildesheim and New York, 1971.
74. Edme Mariotte: Traité de la percussion ou choc des corps. Paris, 1673.
75. Pierre-Louis de Maupertuis: La Figure de la Terre, determinée par les Observations. Paris, 1738.
76. Père Maziere: Les Loix du Choc des Corps à Ressort parfait ou imparfait. [1] (1726), 57 pp.
77. S. Mills: The Collected Letters of Colin MacLaurin. Shiva Publishing, Nantwich, 1982.
78. S. Mills: The controversy between Colin MacLaurin and George Campbell over complex roots, 1728–1729. Arch. Hist. Exact Sci. **28**, 149–164 (1983).
79. S. Mills: Note on the Braikenridge–Maclaurin theorem. Notes and Records Roy. Soc. London **38**, 235–240 (1984).
80. Patrick Murdoch: Mercator's sailing, applied to the true figure of the Earth. With an introduction concerning the discovery and determination of that figure. London, 1741.
81. Sir Isaac Newton: Philosophiae Naturalis Principia Mathematica (First Edition). London, 1687.
82. Sir Isaac Newton: Opticks: or, a Treatise of the Reflexions, Refractions, Inflexions and Colours of Light. London, 1704.
83. Sir Isaac Newton: Optice: sive de Reflexionibus, Refractionibus, Inflexionibus et Coloribus Lucis, Libri Tres (translation by Samuel Clarke). London, 1706.
84. Sir Isaac Newton: Philosophiae Naturalis Principia Mathematica (Second Edition). Cambridge, 1713.
85. Sir Isaac Newton: Philosophiae Naturalis Principia Mathematica (Third Edition). London, 1726.
86. Sir Isaac Newton: Philosophiae Naturalis Principia Mathematica (Jesuit Edition). Geneva, 1739–1742.
87. Sir Isaac Newton: Opticks: or, a Treatise of the Reflexions, Refractions, Inflexions and Colours of Light. Dover Publications, New York, 1952.
88. Sir Isaac Newton: Philosophiae Naturalis Principia Mathematica (First Edition). Culture et Civilisation, Brussels, 1965.
89. A. Partington (editor): The Oxford Dictionary of Quotations. Oxford, 1992.
90. B. Ponting: Mathematics at Aberdeen: Developments, Characters and Events, 1717–1860. Aberdeen University Review, 162–176 (1979).
91. H. Rackham: Cicero, Vol. XVII. Heineman, London, 1971.

92. H. Rackham: Cicero, Vol. XIX. Heineman, London, 1972.
93. Jacques Rohault: Physica. Latine vertit, recensuit, et uberioribus jam Adnotationibus, ex illustrisimi Isaaci Newtoni Philosophia maximam partem haustis, amplificavit et ornavit Samuel Clarke, S. T. P. London, 1710.
94. Jacques Rohault: Rohault's System of Natural Philosophy, illustrated with Dr Samuel Clarke's notes, taken mostly out of Sir Isaac Newton's Philosophy. Garland, New York and London, 1987.
95. R. Schlapp: Colin Maclaurin: a biographical note. Edinburgh Math. Notes **37**, 1–6 (1949).
96. Thomas Simpson: Mathematical Dissertations on a Variety of Physical and Analytical Subjects. London, 1743.
97. James Stirling: Of the figure of the Earth, and the variation of gravity on the surface. Phil. Trans. **39** (No. 438 July–Sept. 1735), 98–105 (1735–36).
98. E. W. Strong: Newton and God. J. Hist. Ideas **13**, 147–167 (1952).
99. Samuel Sturmy: An Account of some Observations, made this present year by Capt. Samuel Sturmy in Hong-road within four miles of Bristol, in Answer to some of the Queries concerning the Tydes, in No. 17 and No. 18. Phil. Trans. No. 41, 813–817 (1668 (1669)).
100. Samuel Sturmy: The Mariner's Magazine (second edition). London, 1679.
101. P. G. Tait: Note on a Singular Passage in the *Principia*. Proc. Roy. Soc. Edinburgh **13**, 72–78 (1884–86).
102. D. P. Thomas: Mathematics Applied to Mechanics. Blackie–Chambers, Glasgow and London, Edinburgh, 1977.
103. I. Todhunter: A History of the Mathematical Theories of Attraction and the Figure of the Earth. Dover Publications, New York, 1962.
104. H. W. Turnbull: Bi-Centenary of the Death of Colin Maclaurin (1698–1746): Mathematician and Philosopher, Professor of Mathematics in Marischal College, Aberdeen (1717–1725). The University Press, Aberdeen, 1951.
105. I. Tweddle: James Stirling: 'This about series and such things'. Scottish Academic Press, Edinburgh, 1988.
106. I. Tweddle: Some results on conic sections in the correspondence between Colin MacLaurin and Robert Simson. Arch. Hist. Exact Sci. **41**, 285–309 (1992).
107. I. Tweddle: The prickly genius – Colin MacLaurin (1698–1746). Math. Gazette **82**, 373–378 (1998) (reproduced in Mathematics Today **37**, 57–59 (2001)).
108. I. Tweddle: Simson on Porisms. Springer-Verlag London, 2000.
109. I. Tweddle: James Stirling's *Methodus Differentialis*. Springer-Verlag London, 2003.
110. C. Tweedie: A study of the life and writings of Colin MacLaurin. Proc. Edinburgh Math. Soc. **8**, 132–151 (1915).
111. C. Tweedie: James Stirling: A Sketch of his Life and Works along with his Scientific Correspondence. Clarendon Press, Oxford, 1922.
112. John Wallis: A Summary Account given by Dr. John Wallis, Of the General Laws of Motion, by way of a letter written by him to the Publisher, and communicated by the R. Society, Novemb. 26 1668. Phil. Trans. III (1668) No. 43 (January 11 1668/9), 864–866 (1668/9).
113. John Wallis: Mechanica, sive, de Motu Tractatus Geometricus. London, 1670–1671.
114. R. V. Wallis and P. J. Wallis: Biobibliography of British Mathematics and its Applications. Epsilon Press, Letchworth, 1986–.
115. D. Weeks: The life and mathematics of George Campbell, F.R.S. Hist. Math. **18**, 328–343 (1991).

116. P.Wood: Science and the Aberdeen Enlightenment. Philosophy and Science in the Scottish Enlightenment (editor: P. Jones) (John Donald Publishers, Edinburgh, 1988), 39–66.
117. P. B. Wood: The Aberdeen Enlightenment: The Arts Curriculum in the Eighteenth Century. Aberdeen University Press, Aberdeen, 1993.
118. Sir Christopher Wren: Dr Christopher Wrens Theory concerning the same Subject; imparted to the R. Society Decemb. 17 last, though entertain'd by the Author divers years ago, and verified by many Experiments, made by Himself and that other excellent Mathematician M. Rook before the said Society, as is attested by many Worthy Members of that Illustrious Body: Lex Naturae de Collisione Corporum'. Phil. Trans. III No. 43 (January 11 1668/9), 867–868 (1668/9). (Reproduced and translated in [53], 319–322).

Index

Sources and Studies in the History of
Mathematics and Physical Sciences

Continued from page ii

Jones A. (Ed.)
Pappus Of Alexandria - Book VII of the Collection

Kheirandish E.
The Arabic Version of Euclid's Optics

Lützen J.
Joseph Liouville 1809-1882

Mcyenn K. von, Hermann A., Weisskopf V.F. (Eds)
Wolfgang Pauli: Scientific Correspondence II: 1930-1939

Meyenn K. von (Ed.)
Wolfgang Pauli: Scientific Correspondence III: 1940-1949

Meyenn K. von (Ed.)
Wolfgang Pauli: Scientific Correspondence IV, Part I: 1950-1952

Meyenn K. von (Ed.)
Wolfgang Pauli: Scientific Correspondence IV, Part II: 1953-1954

Meyenn K. von (Ed.)
Wolfgang Pauli: Scientific Correspondence IV, Part III: 1955-1956

Neugebauer O. (Ed.)
Astronomical Cuneiform Texts

Schubring G.
**Conflicts Between Generalization, Rigor and Intuition: Number Concepts
Underlying the Development of Analysis in 17th-19th Century France and
Germany**

Sesiano J.
**Books IV to VII of Diophantus' Arithmetica: In the Arabic Translation Attributed
to Qustā ibn Luqā**

Sigler L.
**Fibonacci's Liber Abaci: A Translation into Modern English of Leonardo Pisano's
Book of Calculation**

Stedall J.
The Arithmetic of Infinitesimals: John Wallis 1656

Tweddle I.
Simson on Porisms: An Annotated Translation of Robert Simson's Posthumous Treatise on Porisms and Other Items on this Subject

Tweddle I.
James Stirling's *Methodus Differentialis*: An Annotated Translation of Stirling's Text